The Institute of Biology's
Studies in Biology no. 110

Sexual Incompatibility
in Plants

D. Lewis

F.R.S.
Emeritus Professor
University College London

First published 1979
by Edward Arnold (Publishers) Limited
41 Bedford Square, London WC1B 3DQ

ISBN: 0 7131 27473

British Library Cataloguing in Publication Data
Lewis, Dan
    Sexual Incompatibility in Plants. — (Institute of Biology.
    Studies in biology   ISSN 0537–9024).
    1. Plants, Sex in
    I. Title   II. Series
    581.1′6′6   QK827

    ISBN 0–7131–2747–3

Printed in Great Britain by
Thomson Litho Ltd, East Kilbride, Scotland.
Bound by W. H. Ware & Son Ltd,
Clevedon, Avon.

General Preface to the Series

Because it is no longer possible for one textbook to cover the whole field of biology while remaining sufficiently up to date, the Institute of Biology has sponsored this series so that teachers and students can learn about significant developments. The enthusiastic acceptance of 'Studies in Biology' shows that the books are providing authoritative views of biological topics.

The features of the series include the attention given to methods, the selected list of books for further reading and, wherever possible, suggestions for practical work.

Readers' comments will be welcomed by the Education Officer of the Institute.

1979
Institute of Biology
41 Queen's Gate
London SW7 5HU

Preface

I have written this little book in the hope that some of the pleasure and fascination that the study of incompatibility in plants has given me will come through and will be shared by you. The subject is fascinating because it reveals the great refinement at every level, from molecule to behaviour, of biological evolution which ensures that every environmental niche is exploited to perfection. Even the basic means of attaining this perfection by the recombination of deoxribonucleic acid is under control at several levels from the molecule itself to the control of inter-breeding. The different breeding systems in plants of which self-incompatibility is the most important and most suited to their special needs, have been developed in great variety. None of these systems are simple; some are extremely complex, and I have not evaded the difficulties but have tried to reduce technical terms to a minimum so that if you have an elementary knowledge of genetics you should be able to understand the more subtle aspects; otherwise my hope would be that you would understand the more obvious and general conclusions and this in turn would rouse your interest in another fascinating subject – genetics.

London, 1978

Contents

1 Drive for Diversity

'Variety is the spice of life' might have been the first heading and theme of this book, but this implies the trivial and not the essential. Perhaps, 'variety is essential to life' is nearer my meaning, for animals, plants, bacteria and viruses have become what they are and what they will be by the process of evolution through natural selection acting on diverse and varied individuals. For natural selection to be effective, there must be differences between individuals and these differences must be inherited, and therefore they must be genetically determined, or to use a modern concept, they must be encoded in their nucleic acid. Charles Darwin realized the central importance of variation, but explained its preservation by invoking the now discredited hypothesis of the inheritance of acquired characters. Now that we know how inheritance works, we can understand how all organisms from simple viruses to man have developed mechanisms to preserve their variation. Flowering plants are of special interest because they have several different systems of great complexity and subtlety to suit their special needs.

The genetic differences arise in the first place, from sudden random changes in the gene which occur at low frequencies. Single gene mutations may often be effective and useful as immediate sources of variation in bacteria and viruses, where a small colony contains billions of individuals and the time of reproduction is a few minutes; but in higher organisms, the population size is too small, and the life cycle too long for such rare events to provide by themselves a store of variation that can be effective in evolution. A single gene mutation within the genetic background of one individual and one environment is of little evolutionary value; its importance is revealed only when tested in large numbers of individuals and in many environments. Sexual reproduction provides the only effective means of scrambling the genes and their mutations within the imposed limits laid down by the necessity for organisms to have one complete set of genes, if haploid and two such sets, if diploid. Sex does this in two steps; firstly there is a special division of the nucleus, meiosis, preceding the formation of the sex gametes, eggs, sperm, or pollen, which not only halves the number of chromosomes but also rearranges, at random, the two sets of chromosomes which come from the two parents, and also supplies each gamete with a complete set of chromosomes with all the genes they contain. This division also rearranges the genes within a chromosome by the mechanism of crossing-over between parental chromosomes. Secondly, having scrambled the genes into the gametes, the sexual act of mating and fertilization allows the almost random fusion of these gametes to produce virtually unique individuals.

For this elaborate scrambler to work, it must have something to scramble, i.e. there must be normal and mutant genes in the same individual; the individuals must be heterozygous for a proportion of their genes. Heterozygosity can be preserved by the orderly Mendelian segregation of genes, but this requires cross-fertilization between individuals. If self-fertilization occurs, as it can in many plants and some sedentary hermaphrodite animals, such as the oyster, the individuals arising after several generations of selfing are homozygous. Mendel's law of segregation clearly shows how this occurs; if we consider a plant or animal which is heterozygous for one gene, say tall and dwarf in Mendel's pea plant, the offspring after one generation of selfing will be: one tall TT homozygote, one dwarf tt homozygote, and two tall Tt heterozygotes. One generation of selfing has reduced the heterozygotes to one half. If the original plant had been heterozygous for ten genes, then the offspring from selfing, on average, would be heterozygous for only five genes.

Animals, having their sexes in different individuals, cannot self, but they can cross-fertilize between brother and sister. This produces a similar inbreeding effect to selfing but at a slower rate. Cousin mating also produces the same effect but at a still slower rate. Whether it be selfing, brother-sister or cousin mating, the long-term end result is homozygosity, true breeding and little or no variation to scramble.

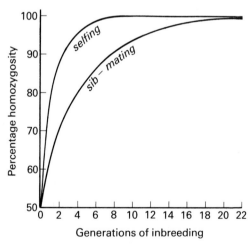

Fig. 1–1   The relationship between the number of generations of selfing and sib-mating and the % of homozygous genes.

Figure 1–1 shows the effects over twenty years of two types of inbreeding.

Hermaphrodite plants, for continued evolution, must be able to control self-fertilization by preventing it or reducing it to a low level, but as a temporary measure it may be necessary to resort to self-fertilization for survival under difficult conditions which might prevent cross-pollination. Self-fertilization is not a problem in animals, and brother-sister (sibling) mating is solely controlled by behavioural instincts and taboos. Sib-fertilization in plants has been considered by some authors to be of little importance, but once self-fertilization has been controlled, it can be argued from the limited dispersal of seed and pollen and from the efficient incompatibility systems which do limit sib-fertilization, that it is worthy of consideration. Many ingenious experiments have been carried out with both wind- and insect-pollinated plants to measure the distribution of pollen in distance from the source. They show a remarkable general agreement, in that more than 90% of the pollen is deposited within 1–4 metres of the source, and a small percentage, diminishing with the distance, is deposited over many kilometres. Seeds are similarly restricted.

These two facts are very important factors in the population structure and mating of plants. Most of a plant's close neighbours will be its sibs, most of the pollen that arrives on a flower will be from sibs. This close relatedness of pollen and seeds is affected by other factors which are difficult to assess; the density of plants on the ground and the number of flowers open at any time on the plant are the obvious ones. Wild cherry trees may be further apart than buttercup plants, but the higher number of flowers on the cherry counteracts the bigger distance. Plants, as adults, are static, and their pollen and seeds are carried by wind or animals. The whole economy of sexual reproduction depends upon the intricate adaptation between the plant and the transporting agents, and by these adaptations the plant has not only to control its breeding system but also to reproduce economically without excess wastage of pollen and eggs. The way plants have overcome their limitations is the main subject of this book, but in order to get the full meaning we must consider briefly why the system of separation of the sexes which is so highly successful in animals, has been tested in plants and generally rejected.

About 5% of the flowering plants have male and female sexes in different individuals, the hop, asparagus and cannabis providing examples. These dioecious species occur sporadically throughout the plant kingdom and have evolved from the common hermaphrodite plants. They have not evolved into large and successful groups. Separation of the male and female in separate plants is probably extremely wasteful, for not only does half the population produce no seeds, but the pollen potential must exceed the requirements many times. This separation of the sexes does, of course, exclude the most extreme form of inbreeding, selfing, but it offers no protection against sib-mating, because pollen from a male plant is equally effective on a female whether it be a sib or non-sib. If, as I believe, the control of sib-fertilization is important, separation of the sexes has a limited value as an outbreeding device in static plants. Animals, with their mobility, behaviour, perception and communication have developed the separation of the sexes to perfection. The vagaries of relying on wind or erratic insects has been replaced by a highly motivated and efficient transfer of sperm by coitus. This provides the necessary economy in reproduction. Animals have instinctive migration behaviours, while man, in addition, has conscious traditions and taboos to control fertilization; plants without mobility, motivation, or consciousness have developed a genetic system of equal or greater efficiency.

2 The Promiscuous World of Pollination

Flowering plants have three means of transferring male pollen to female stigma, wind, water and animals. Wind and water were here before the flowering plants evolved; several groups of insects preceded the flowering plants by many millions of years, and these insects, beetles, thrips, sawflies, are still active but inefficient pollinators of flowers. The insects which are specialized for flower feeding and pollination are the *Hymenoptera* (bees and wasps) and the *Lepidoptera* (butterflies and moths) and these appeared at the same time as the flowering plants. We should remember that the plant has two seemingly opposite requirements from pollination: (i) to restrict pollen to the individuals of one species and not waste it on other species, and (ii) to discourage pollen from reaching the stigma of too nearly related individuals of its own species.

If pollination is by wind or promiscuous insects, little can be done by the plant to direct the pollen to the right stigma. But the fact that most temperate trees, as well as all conifers and grasses together with several other groups of plants are wind-pollinated and are still flourishing, shows that the problem has been solved well enough at best to compete with plants that are more efficiently pollinated by bees. It has been achieved by the production of large amounts of light pollen in exposed and wind-swept anthers. The female stigma is large and feathery, offering large trapping surfaces, and the ovary contains one or very few seeds, so that if only two to three pollen grains of the right type land on the stigma, there will be full fertility. Contrast this with a poppy flower with >1000 seeds which therefore requires a few thousand pollen grains, and this can only be obtained by species-constant pollination. Many species of bees, butterflies and moths are extremely species-constant in their visits; this behaviour has been found by direct observation of marked bees, by examination of the pollen loads of bees, and also from the contents of combs in the hive. Parallel with the evolution of constant insects, the plant has reinforced this by making their flowers distinct in shape, pattern, scent, and, to a lesser extent, in colour. Flowers have also evolved intricate structures which present difficulties in the extraction of pollen or nectar, so that a bee becomes a specialist and more set in its habits. The snapdragon requires a bee having enough strength to open the mouth of the corolla tube. The nectar in some clover species can only be tapped by a species of bee with a long proboscis. Species which have these mechanically closed flowers are invariably pollinated by constant insects – bees, butterflies or moths. The more open type of flower as found in *Ranunculus* are more often pollinated by promiscuous insects. At the same time, flowering plants have evolved both mechanical, temporal and biochemical devices to reduce self-pollination. Many hermaphrodite plants have their sexes in different flowers on the same plant; some plants mature their pollen and eggs at different times. The maize plant is a good example of the effectiveness of temporal and spatial separation of the sexes. The male tassel at the tip of the plant matures its pollen at least a day before the female cobs, which are in the axil of a lower leaf, are receptive. The maize plant only produces one tassel and one or two cobs. Furthermore, the pollen, like much wind-borne pollen is short lived. The combination of all these factors make it a very efficient restriction to self-pollination, but of course, only in plants which have a small number of flowers. It is not the intention of this book to give a full account of these methods. But even if these mechanical devices can in some plants reduce self-pollination, they have no effect on sib-pollination, a type of inbreeding which, because of the extremely narrow distribution of pollen and seeds as we have seen earlier, is of considerable importance. Only genetical-biochemical methods of self-incompatibility can effectively control this, not by controlling pollination, but by controlling fertilization.

3 The Drive Against Inbreeding

The majority of flowering plants, being hermaphrodite, produce fertile male spores (pollen) and fertile female gametes (eggs) but many species are unable to reproduce sexually by self-pollination; they are *self-incompatible*. A typical example of the scope and efficiency of one type of self-incompatibility can be found in a field of clover: every plant is self-incompatible but to find a pair of plants which were cross-incompatible would require testing on average more than 22 000 pairs. Not all self-incompatible plants have such an efficient system as the one in clover. There are five main systems known and probably there exist others which have not yet been discovered. They all work either by a specific inhibition of pollen penetration of the stigma, or of pollen tube growth in the style, both of which prevent the male nucleus fertilizing the egg; or more rarely, as in cacoa, the egg is fertilized, but early abortion occurs after selfing. There are several logical ways of classifying the systems, but for convenience we will use one based upon morphology; these are *Heteromorphic* in which differences in flower morphology characterize the inter-compatible types and *Homomorphic,* in which there are no such differences.

3.1 Heteromorphic incompatibility

3.1.1. *Primrose*

The common European primrose, *Primula vulgaris*, is a good example to illustrate the main features of the heteromorphic system. If you examine the flowers of populations of primroses, except some certain rare populations in Somerset, England, which will be described later, you will find that about half of the plants are long styled, also called pin-eyed flowers, with the rounded stigma at the mouth of the corolla tube and the anthers attached to the tube at about the mid-point; the other half of the plants are short-styled, or thrum-eyed, in which the anthers are at the mouth of the tube and the stigma is at the mid-point. Figure 3–1 shows these two types.

There are other morphological differences which require a micro-scope to reveal. The pollen grains in the short-styled plant are larger than those in the long-styled plant, they have about twice the volume and, as we shall see later, they have twice the distance to grow to achieve fertilization. The surface sculpturing of the pollen grains is also distinct. The surface of the stigma of the long-styled plant has larger cells than those of the short-styled. These morphological differences help, but not very effectively, to distribute the pollen from one type to

Fig. 3–1 (a) Long-styled (pin-eyed) and (b) short-styled (thrum-eyed) flowers of *Primula obconica.*

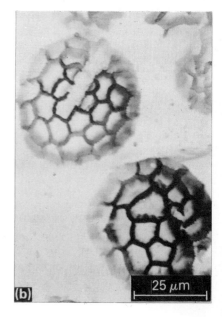

Fig. 3–2 Scanning electron micrographs of stigmas and pollen in *Limonium meyeri*; by permission of D. R. Dulberger and the Royal Society. (**a**) Stigma of long style. (**b**) and (**c**) Pollen of short-styled plant. (**d**) Stigma of short style. (**e**) and (**f**) Pollen of long-styled plant.

the other. The corresponding height of anthers and stigmas between the two types and the restricted diameter of the corolla tube helps to effect cross-pollination. The size of the stigmatic cells and the pollen sculpturing may help to assist the adhesion of the pollen on the stigma of the other type. These pollen and stigma surfaces are well developed in the heteromorphic species of sea lavender, *Limonium myerii*.

But these morphological differences give only fringe benefits of doubtful value and are the outward and secondary trappings of a much more important difference which is the basis of incompatibility. If the pollen of either the long- or short-styled plants is placed on its own type of stigma, whether it be from the same plant or another plant of the same type, the pollen tube will penetrate the stigma and will grow to 1 or 2 mm into the style and then stop growing, thus preventing the nucleus reaching the egg and hence producing no seeds. This was called by Darwin 'an illegitimate union'. The legitimate or compatible union is between the two types in which the pollen tubes grow at a steady rate, reach the ovary in 18–24 hours and produce seed (Fig. 3–3). The biochemical nature of the inhibition in

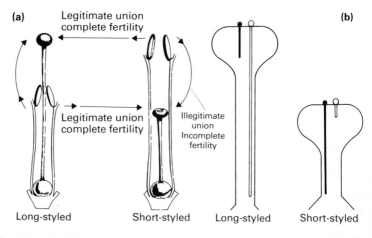

Fig. 3–3 (a) The two forms of *Primula* indicating the two legitimate (compatible) and two illegitimate (incompatible) pollinations (from DARWIN, 1877); (b) the pollen tube growth from the four types of pollination, note the difference in pollen size. ●, pollen of long-styled plant; ○, pollen of short-styled plant.

the incompatibile union is discussed later, but it should now be realized that it is a process of self-recognition which results in rejection of self. The consequences in terms of the breeding system and out-breeding are identical to the effect of separation of the sexes in different individuals as in animals and a few plants. The long- and short-styled types in *Primula* and other distylic species and the two sexes in dioecious species are both in equal proportion in the population. This equality is ensured by the genetic control of sex which is by the two X-chromosomes in the female and an X- and Y-chromosome in the male. In distylic species, the genetic control is by a cluster of genes designated S and s; the long-styled plant is homozygous ss, and the short-styled plant is heterozygous Ss. When they are crossed together in the compatible combination they produce long- and short-styled plants in equal numbers:

Parents Long × Short
ss Ss
↓
Progeny 1 Long : 1 Short
ss Ss

Both distyly and separation of the sexes prevent self-fertilization and fertilization between individuals of the same groups or sex, but neither give any restraint or control on sib-fertilization. This is because the proportion of the two groups is the same in a family and the population as a whole. The major difference between separation of the sexes and distyly is that both types in distyly produce offspring and hence there is a considerable crude economy in reproduction.

Heteromorphic incompatibility is found in approximately 17 families and about 100 genera (see Appendix). The full complex of morphological differences found in the primrose is found in several other genera, including the sea lavender, the pollen and stigmatic surfaces of which are illustrated in Fig. 3–2. Some species, for example, *Linum grandiflorum*, have styles of different lengths, but the anthers are at the same height in the two types and the pollen is of the same size, although the two pollen types differ in their osmotic pressure, which is another possible way to compensate for the different style lengths through which the pollen tubes have to grow. The one common feature of heteromorphic species is the incompatibility.

A comparative study of species and genera in the sea lavender family, the *Plumbaginaceae*, shows that heteromorphic incompatibility has evolved in a sequence from monomorphic species which are self-incompatible through species with dimorphic pollen, dimorphic pollen and stigmatic cells, to the full heterostyly. This indicates that incompatibility predates heteromorphism.

An important feature of self-incompatibility systems is that they can change to self-compatibility without complete disruption of sexual reproduction, and if for some change in the environment self-compatibility becomes an advantage, or even a necessity, then such a change would be selected. For example, if the particular insect pollinator was not active, survival would depend upon the presence of the rare self-compatible plant. Self-compatible plants do exist in heteromorphic species and they are usually homostyles, i.e. they have the style and stigma of one type and the anther and pollen of the other. These are known as cultivated varieties of several species, such as *Primula hortensis, P. obconica* and *P. viscosa*; they probably arose as rare types which were unconsciously selected by growers for their self-compatibility. There are several populations of the primrose *P. vulgaris* which contain 50–60% of self-compatible long homostyles found in Somerset, England. There is no obvious reason for these local

Fig. 3–4   Long homostyle flower of *Pemphis acidula* from Malagasy. (**a**) General surface view; note that the stigmas and anthers are both at the top of the tube. (**b**) Half-section through flower, the anthers have shrunk in the preparation and in the fresh flower are on the same level as the stigma.

populations of homostyle primroses because they occur sporadically and within an area which contains the normal long- and short-styled populations. Perhaps a careful study of the local ecosystems might reveal a correlation.

In *Pemphis acidula* (Fig. 3–4), a plant which only grows in the tropics on coral at sea level and often on isolated islands, most of the populations are typically distylic, but in southern Malagasy the population is composed of 100% long homostyles that are self-compatible. The completely insular environment of this population, and some local situation affecting the normal pollinating insect, or a small founder population following one of the devastating changes in sea level, are possible explanations for this unique population. These homostyle types are the result of a rare recombination by a cross-over between the genes in a cluster, as we shall see in a later chapter. Only if the crossing-over occurs at a point which separates and recombines all the pollen/anther characters from the style characters is the plant self-compatible. Long homostyles are known which have high anthers but small pollen, and these are self-incompatible.

3.1.2. Purple loosestrife

Heteromorphic incompatibility with its two interbreeding types is not as efficient, both in the economic use of pollen and in the control of sib-mating, as other systems of incompatibility, and it is of interest to note how the accumulation of the morphological differences have hindered the development of the more efficient outbreeding systems found in plants such as in red clover, which have not two but thousands of interbreeding types. It is obvious that the number of discrete types that could be obtained in the lengths of styles and the height of anthers is severely limited; evolution has, however, extended the system to three types, trimorphy, in three families, the *Lythraceae, Oxalidaceae* and *Pontederiaceae.*

The common loosestrife, *Lythrum salicaria,* shows all the features of the trimorphic system. Populations, which can be seen along canal, river and dyke banks in Europe, contain long-, short- and mid-styled plants in approximately equal proportions (Fig. 3–5). There are three levels in each type of flower; one level is occupied by the stigma and the other two by two whorls of anthers. The two levels of anthers in any one type correspond to the levels of stigmas in the other two types.

There are three different sizes of pollen, the largest is in the high anther, the smallest in the low anther and an intermediate in the mid anther. All pollinations, six in number, from anthers to stigmas at the same level are compatible, and all at different levels, twelve in number, are incompatible. The rate and final extent of compatible pollen tube growth is correlated with pollen size. The growing pollen tube exerts a stimulus on the developing ovary so that the time, after pollination, of the arrival of the tube in the ovary is critical for fertilization. Because of the difference in pollen size the tube reaches the ovary in the same time despite the wide disparity in style length. The genetic control of trimorphism is by separate genes s and m the long-styled is a recessive homozygote for both genes $mm\ ss$; the mid-styled is $Mm\ ss$ and the short-styled is $mm\ Ss$ or $Mm\ Ss$. As in dimorphism, the short-style is inherited as a dominant character.

From a superficial view it might be thought that trimorphic incompatibility has advantages over dimorphy because of the increase of mating types from two to three, but a closer look reveals that it is

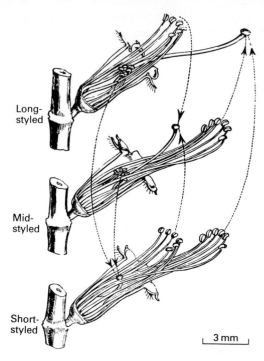

Fig. 3–5 Diagram of flowers of the long, mid- and short-styled forms of *Lythrum salicaria*. The dotted lines with the arrows show the directions in which pollen must be carried to each stigma to ensure compatibility. (From DARWIN, 1877.)

a mixed blessing. Both systems control self-fertilization and fertilization within a type; sib-fertilization is not controlled at all in dimorphy and only to a small extent in trimorphy. We can make an estimate of sib-compatibility in trimorphy if we make the reasonable assumption that 75% of the short-styled plants have the genotype *mm Ss* and 25% *Mm Ss* the long-styled are all *mm ss* and the majority of mid-styled are *Mm ss* although *MM ss* do exist. If we then consider all the possible compatible crosses in the population, we can calculate the number of crosses such as Long × Short, *mm SS × mm Ss*, which give two groups, long and short in the offspring; and also the number of crosses, such as Mid × Short, *Mm ss × mm Ss*, which give three groups, long, short and mid in the progeny. We can also calculate the percentage of compatible pollination with a 'two-group' and a 'three-group' family. It is 25% in a two-group family, because only two out of the eight different pollinations between any two groups are compatible. The percentage in a three-group family is 33% because six out of the eighteen pollinations are compatible. Taking all the families, there are seven two-group families, and five three-group. The mean compatibility of the total of the families is

$$\frac{(25 \times 7) + (33.3 \times 5)}{12} = 28.4\%.$$

The general compatibility in the population as a whole is 33.3% as with a three-group family. The ratio of sib to general compatibility is $28.4/33.3 = 0.85$. The corresponding ratio for dimorphy is 1.00, so there is a slight reduction of sib-fertilization in trimorphy. This slight advantage is more than offset by an increase in the wastage of pollen. If pollination is at random, 50% of pollen is wasted in dimorphy; in trimorphy with the 33.3% of compatible pollination there is 66.6% of pollen wastage.

This pollen wastage is the result of having two different types of pollen in the same plant, and has laid a heavy burden on the development of trimorphy. This fact, and the limitations of effectively dividing the flower parts into four divisions, stops any further development in heteromorphic incompatibility. It is of interest to note that trimorphism has evolved in only three families and all the families have two developmentally distinct whorls of anthers, episepalous and epipetalous. It would appear that the contrasting differentiation of the anther length, pollen size and the incompatibility reaction in the same flower are only possible where two distinct whorls are present. In view of the doubtful advantage of trimorphy, and the greater wastage of pollen and the fact that dividing the levels into three instead of two cannot be as efficient in the distribution of pollen to the right compatible place, it is surprising that trimorphy ever evolved, and we must look for other hidden reasons for its evolution. It is significant that there are several dimorphic species in the trimorphic families, and there are trimorphic species of which large populations have only long- and short-styled plants. It is generally agreed that the dimorphic species in these families have been derived by loss of the mid-type from a trimorphic ancestor.

The coral island plant *Pemphis acidula,* which is a member of the *Lythraceae,* is today dimorphic, but it shows differences in anther heights and pollen size which are remnants of a trimorphic origin. To make all the pollen effective, there has been a mutation which alters the incompatibility of the pollen in the mid-position, so that it is now compatible on the other type.

3.2   The peculiar genetics of pollen grains

The confronting elements of incompatibility are haploid pollen grains and diploid stigmas and styles. The stigma and style are genetically similar to other tissues of the plant. They have two sets of chromosomes and genes and their characters show all the features of

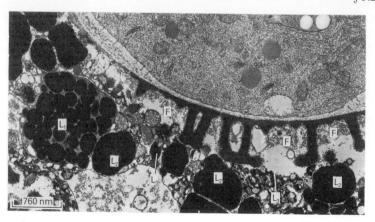

Fig. 3–6 Section of pollen grain of *Raphanus sativus* showing fibrillar material, F, and lipoprotein. L_1, L_2, L_3.

diploid inheritance including dominance and recessiveness, in which one allele is active, and codominance in which both alleles are active. The pollen grain, however, is haploid with one set of genes, and shows for many characters haploid genetics in which dominance cannot be expressed. The pollen grain is peculiar in several respects; it is a differentiated cell of the diploid mother plant, the sporophyte, and has a haploid nucleus and an independent existence. Some of the structures of the pollen grain are derived directly from the sporophyte and show maternal diploid inheritance. Other characters and structures formed later are under the control of the haploid nucleus and show haploid inheritance. When the pollen grain is nearly mature, lipoproteins which are synthesized in the tapetum layer of the sporophyte are poured over it. This complex situation was first realized from the genetics of pollen characters. The first pollen character to be analysed was the shape of the pollen grain in the sweet pea in 1905 by Bateson, Saunders and Punnett. Long pollen, L, was found to be dominant to round, l. The pollen grains of the hybrid plant which is heterozygous, $L\,l$, produces 50% of the pollen grains with L and 50% with l but all these pollen grains are long. Round pollen grains are only produced by $l\,l$ homozygous plants and then all the pollen is round. The character, long or round, is determined *before* the genes L and l segregate at meiosis, and we say the character is maternally, or *sporophytically* determined.

The outpouring of lipoprotein onto the pollen grain has its origin in sporophytic tissue and this too is sporophytically determined.

Early workers on self-incompatibility assumed that all pollen characters were sporophytically controlled and were in difficulties explaining their results until they considered gametophytic determina-

Fig. 3–7 Pollen grains from a single anther on a stigma of *Oenothera organensis*. Half of the pollen grains are empty and compatible and have produced the pollen tubes in the stigma (on the right), and half the grains are full and incompatible. This is an example of gametophytic control of the pollen incompatibility.

tion in 1924. This was when the character starchy *Wx* and waxy *wx* pollen in maize was described. The pollen from a hybrid maize plant heterozygous for *Wx wx* segregates into 50% starchy and 50% waxy. This is revealed directly by staining the pollen grains from a single anther, with a solution of iodine, when 50% are stained dark blue and 50% brown. In this example, the gene *Wx* produces its effects *after* segregation at meiosis, and the character is conditioned by the pollen grain's own gene, hence it is *gametophytically* determined (see Fig. 3–7 for an example).

This distinction between a gametophytically and a sporophytically determined character is crucial for the understanding of incompatibility systems. Heteromorphic systems have their relevant pollen characters, size, sculpturing and incompatibility reaction determined sporophytically. The pollen of a short-styled *Primula* plant, which is heterozygous for *Ss* will comprise 50% with *S* and 50% with *s*, but all the pollen is large, has the same sculpturing and the same incompatibility reaction. If this were not so, half the pollen would be compatible on its own style. For self-incompatibility in heteromorphy, it is essential that the pollen should be sporophytically determined, and that there should be dominance of *S* to *s* in pollen and style.

3.3 Homomorphic gametophytic clover, tobacco, cherry

The sweet cherry, *Prunus avium*, ornamental tobacco, *Nicotiana alata*, clover, *Trifolium pratense,* and many other plants distributed in 56 families (see Appendix), are self-incompatible but they show no visible differences in the flower. The incompatibility is said to be homo-morphic. If two plants are crossed and the resulting family is tested for cross-compatibility between all the progeny, either one of two different types of compatibility pattern is obtained. One type has two groups of approximately equal numbers of individuals and the other type has four groups; all cross pollinations between plants in the same group are incompatible, but pollinations between any two plants taken from different groups are compatible. Figure 3–8 shows a mating matrix of two idealized families.

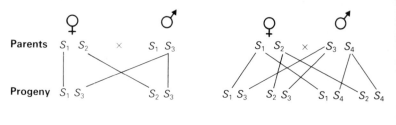

Fig. 3–8 The two types of compatible crosses in the homomorphic gametophytic system. Above are the genotypes of the parents and progeny. Below are mating reactions between the progeny; +, compatible, −, incompatible.

This can be contrasted with dimorphic *Primula* which regularly produces two-group families. The explanation for the two families lies in the nature of the controlling gene and the gametophytic control of the pollen reaction. The controlling gene S is really a cluster of genes as we shall see later, but for the present analysis it segregates as a single unit and can be treated as a single gene. This gene exists in a large series of different alleles, these are designated $S_1, S_2, S_3 \ldots S_n$. Each diploid plant has two *different* alleles for reasons that will become obvious. In the formation of the pollen, the two different S alleles separate, 50% of the pollen receiving one allele and 50% the other. The incompatibility reaction of these alleles causes a pollen tube to be inhibited in a style containing the same allele as the pollen grain. The self- and different cross-pollinations are illustrated in Fig. 3–9.

For a cross-pollination to be incompatible, the two parental plants have the same two alleles. If the parental plants have only one allele in common, then only half the pollen functions and a two-group family is produced because of the segregation of the S alleles in the eggs of the female parent. If there are no alleles in common, then both types of pollen function, and meeting two different types of eggs, produce a four-group family. There is no dominance of alleles in the style, both are operating to oppose the growth of the respective pollen tube; the

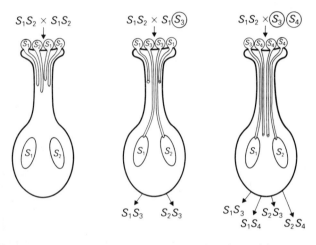

Fig. 3–9 Diagram showing styles and pollen of a plant with gametophytic incompatibility; on the left is an incompatible pollination, the one in the centre is compatible giving two groups, and the one on the left is also compatible, and giving four groups in the family. (After CRANE, M. B. and LAWRENCE, W. J. C. (1929). *J. Pomol.*, **7**.)

alleles are *codominant*. Codominance in the style is essential with gametophytic control because if one allele was dominant, half the pollen in a self-pollination would have no active allele to oppose it, and the plant would be self-compatible. Comparing this with the dimorphic system with sporophytic control and dominance shows how delicately are the components of these systems coordinated, and because there are hundreds of different alleles, and pollen has one allele and the style two different ones, we can deduce something fundamental about the way the S gene interferes with the pollen tube growth. The basic question is whether the growth interference is due to lack of some necessary stimulus or growth factor, or is it due to a positive inhibition. One could imagine that two *different* alleles in pollen and style were necessary for growth of the tube by the complementary action between them, or we might suppose that where the *same* allele is present in pollen and style, a reaction occurs between their products to trigger the synthesis of an inhibitor, or both might interact at the membrane surface to block it. These two basic possibilities can be summarized as the *complementary stimulus* between different alleles, and the *oppositional inhibitor* between the same allele. We can look at the situation diagrammatically (Fig. 3–10).

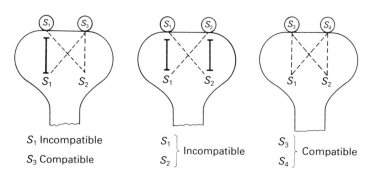

S_1 Incompatible

S_3 Compatible

$\left.\begin{array}{l} S_1 \\ S_2 \end{array}\right\}$ Incompatible

$\left.\begin{array}{l} S_3 \\ S_4 \end{array}\right\}$ Compatible

Fig. 3–10   Diagram of pollen and styles to show the difference between an oppositional inhibitor mechanism (————) between the *same* alleles and a complementary stimulus (------) between *different* alleles.

It is clear from this that if a complementary system were at work, all self-pollination would be compatible as shown by the broken 'complementary lines'. The conclusion from codominance and gametophytic control is that the action is an oppositional inhibitor between the same alleles.

The homomorphic incompatibility is an outbreeding system of great efficiency, both in respect of pollen usage and of control of sib-mating. Because there are hundreds of different S alleles and thousands of different pairwise combinations in diploid plants, the general compatibility of the pollen in a large population is virtually 100%. This gives the economical use of pollen. The four-group family from a cross, which would in nature be the most frequent, gives 25% of incompatible crosses, (see Fig. 3–8), but on a pollen basis, which is the true measure, the percentage of sib-incompatible pollen is 50%. With a two-group family, the sib-compatible pollen is 25%. This gives a value of between 0.5 and 0.25 for sib-compatibility which is a highly efficient control of sib-mating.

One important characteristic feature of the gametophytic system is that the inhibition is on the pollen tube after it has penetrated the stigma and style, and not an inhibition of pollen germination or penetration of the stigma.

3.4 Two-gene system in grasses

A variation on the gametophytic system is found in the grasses. In all species that have been investigated, there are two unlinked genes, S and Z, which control incompatibility. Both these genes have a series of alleles and their cooperative action is extremely interesting. An incompatible reaction to inhibit the pollen tube growth occurs only between pollen grains and styles which have the same S and the same Z alleles. In other words, both S and Z alleles must be matched in pollen and style to cause inhibition. A pollen grain has a unique incompatibility recognition which is based upon the unique combination of a particular S allele with a particular Z allele. If there were 100 different S alleles and 100 different Z alleles, there would be 10^4 different pollen reactions. The two genes segregate independently, and this increases the number of groups in a family; for example, if the two parents differ in both alleles at S and Z, for example, $S_{1,2}Z_{3,4} \times S_{5,6}Z_{7,8}$, then there will be 16 different genotypes and this will be expressed as 16 different sib-breeding groups. But also depending upon the number of alleles common to the two parents, families can occur with 12, 8, 4, 3 and 2 groups. To understand the production of a two-group family, we must realize that, because both S and Z alleles must be matched, plants which are homozygous for one gene such as $S_{1,1}Z_{3,4}$ are obtained from a compatible pollination and, by crossing this plant with $S_{1,1}Z_{3,5}$, two genotypes $S_{1,1}Z_{3,5}$ and $S_{1,1}Z_{4,5}$ are obtained. Most of these families have been found in rye, *Secale cereale*, and meadow grass, *Festuca pratensis*.

Apart from what seems to be an unnecessary complication of two genes, the system is less efficient in the control of sib-mating; the sib/general compatibility ratio is 0.5–0.75 with two genes, as compared with 0.25–0.5 with one gene. The efficiency in the use of pollen in the population as a whole is not significantly different. One feature of the two-gene system which could affect pollen use, is the fact that the number of effective incompatibility specificities is the product of the number of S and Z alleles. There have been found 12 S and 14 Z in *Festuca pratensis*; these generate 168 different specificities, which is equivalent to 168 different alleles of a one-gene system. This is not a particularly high number, for estimates of approximately 400 have been found with one gene. In a later chapter, it will be argued that there may be a restriction imposed by molecular constraints on the number of workable alleles in a two-gene system which is not present with one gene. It is suggestive that the grasses are monocotyledons, which are considered to be primitive; a recent study of incompatibility in two primitive families in the dicotyledons, the *Ranunculaceae* and *Chenopodiaceae* have revealed three and four genes. This strengthens the view that the multi-gene systems are primitive, and the one-gene system is an advanced derivative.

3.5   Sporophytic cabbage, radish and sunflower

It may be no more than a coincidence that the most complicated incompatibility system is found in what are considered by taxonomists to be the most advanced families, the *Compositae* and *Cruciferae*. The system is homomorphic, but the pollen reaction is controlled by the sporophyte. Its complication arises from the large number of alleles as in the gametophytic system, and a complex interlocking hierarchial and parallel series of dominance and codominance between the alleles. Experimental analysis has only touched the fringe of the complexity, and only where the work has had a useful application to plant breeding as in Brussels sprouts and cabbages, which we will describe later, have the action of more than five or six alleles been analysed. However, probably all the basic variations on the dominance theme have been found. As an illustration of these variations, we can consider four alleles, $S_{1,2,3 \text{ and } 4}$; we must indicate the dominance relationship of these alleles, and this is usually shown thus, $S_{\dot{1},3}$, the dot over the figure shows dominance of that allele in the pollen, the line underneath indicates dominance in the style. In the example above, S_1 is dominant over S_3 both in pollen, or more strictly, in the pollen mother cell, and in the style. In the pair, $S_{\dot{1},\dot{2}}$, both alleles are codominant and both are expressed in pollen and style. The incompatibility reaction occurs between pollen grain and style which have the same active allele. For example, all pollen of an $S_{\dot{1},3}$ plant is inhibited only on stigmas containing an active S_1. All pollen of an $S_{\dot{1},\dot{2}}$ will be inhibited on a style with $S_{\dot{1}}$ and on a style with $S_{\underline{2}}$ if S_2 is dominant, e.g. $S_{\underline{2},3}$.

Let us consider the four alleles in their six pairwise combinations: $\dot{\underline{1}},2$; $\dot{\underline{1}},3$; $\dot{\underline{1}},4$; $\dot{\underline{2}},3$; $\dot{\underline{2}},4$; $\dot{\underline{3}},4$. The dot above and the line below indicate the dominance in pollen and style. By crossing these in different combinations, we can obtain a four-, a three- and a two-group family. From the cross $\dot{\underline{1}},3 \times \dot{\underline{2}},4$ we get the four genotypes: $\dot{\underline{1}},2$; $\dot{\underline{1}},4$; $\dot{\underline{2}},3$; and $\dot{\underline{3}},4$, and these have four different mating patterns, ((**a**) below); from $\underline{2},4 \times \dot{\underline{3}},4$ again four different genotypes are produced, but two of these have the same mating reactions because of dominance, and then we have a three-group family with one group having twice as many plants as the others, ((**b**) below); from the cross, $\dot{\underline{1}},3 \times \dot{\underline{3}},4$ we again get four genotypes but these fall into two indistinguishable pairs, and a two-group family results, ((**c**) below).

In all the families given, the parents are reciprocally compatible or incompatible, whichever parent is used as the male or female, and this is similar to the rule in the gametophytic system, but it is not always the

(a)

♀＼♂	1, 2	1, 4	2, 3	3, 4
1, 2	−	−	−	+
1, 4	−	−	+	+
2, 3	−	+	−	+
3, 4	+	+	+	−

(b)

♀＼♂	2, 3	2, 4	3, 4	4, 4
2, 3	−	−	+	+
2, 4	−	−	+	+
3, 4	+	+	−	+
4, 4	+	+	+	−

(c)

♀＼♂	1, 3	1, 4	3, 4	3, 3
1, 3	−	−	+	+
1, 4	−	−	+	+
3, 4	+	+	−	−
3, 3	+	+	−	−

case in the sporophytic system. For example, if we consider another two alleles, 5 and 6 in the pair $4,5$, 4 is dominant in pollen and style, but 5 is dominant only in the style. If $4,5$ is crossed with $5,6$ as male, the cross is incompatible because 5 is active in both pollen and style; but in the reverse cross the 5 is not active in the pollen and therefore it is compatible on the style of $4,5$. These reciprocal differences and the three-group family distinguish the sporophytic from the gametophytic system. One other important point is that S homozygotes are a regular and legitimate part of the system; in the two-group family, one quarter of the progeny are 3,3 and in the three-group family, one quarter are 4,4. These are produced because sporophytic control allows a pollen grain carrying a recessive allele to function although the same allele is present in the style. From a cross between two homozygotes, $S_{3,3} \times S_{4,4}$, a one-group family of $S_{3,4}$ plants is produced, all individuals of which are self- and cross-incompatible. We shall see later that this feature has been of great value in the utilization of self-compatibility in the creation of better vegetables and fodder plants by making F_1 hybrids on a commercial scale.

The sporophytic system is efficient in the economical use of pollen because of the large number of alleles, which means that the majority of random cross-pollinations within a population are compatible. Sib-mating is also efficiently controlled. In a family in which the four genotypes show codominance, 75% of the sib-crosses are incompatible. These types of family are probably not uncommon because from our best estimate in *Brassica oleraceae*, the number of allele pairs which have codominance is 61%. The family with the lowest sib-incompatibility is a two-group family with 50%. The sib/general compatibility is therefore between 0.25–0.5.

The fact that two entirely different and highly efficient systems, the sporophytic and the gametophytic, have evolved in plants points to the great importance outbreeding has for plants. There are fundamental cytological differences in the pollen of the two types, which are characteristic of the taxonomic families, and there is a complete correlated difference with the place of inhibition. The sporophytic inhibition is always on pollen germination and penetration of the stigma. It may be that these long-established cytological differences have dictated the direction of evolution, gametophytic or sporophytic.

As with the gametophytic system, three- and four-gene systems have recently been found in the sporophytic system. The tropical oil seed plant, *Eruca sativa*, has a 3–4-gene system, and this will be discussed later in relation to the evolution of incompatibility systems.

3.6   Sporophytic-gametophytic system, or self-abortion in cacao

In all the systems we have described the incompatibility is a contraceptive device to prevent fertilization. Even if the incompatible pollen is on the stigma many hours before compatible pollen arrives, the eggs are still receptive for later fertilization. This is probably one of the most important features of the systems for it conserves eggs from spoilage by the promiscuity of unwanted pollen. The cacao plant, *Theobroma cacao*, is an oddity in this respect. Most of the high quality cacaos are self-incompatible, the incompatible pollen germinates normally, and the tube grows down the style at full compatible rate, the pollen nuclei enters the embryo sac, fertilizes the egg and the endosperm undergoes a few nuclear divisions and embryo and endosperm abort. The eggs are ruined for compatible fertilization even if the pollen is applied only some thirty minutes after the incompatible pollen. This is a high price to pay for outbreeding. Cacao is unusual not only in the mode of blocking self-fertilization, but also in its genetical control which is gametophytic for the pollen, and sporophytic for the eggs.

4 The Genes in Control

We have referred to one gene, S, being in control of distylic and two genes, S and M, in control of tristylic incompatibility, and similarly S and Z genes in homomorphic incompatibility. We have referred to them as genes because they segregate as single units in inheritance. This is one of the basic features of a gene; they give 1:1, 3:1 and 1:2:1 ratios. But we know that genes each produce *one* limited product, an effect which has been summarized in the aphorism, 'one gene, one enzyme'. The differences between a long- and short-styled plant are many, and are expressed in different tissues and cells such as pollen grains, style and anther filaments. It is inconceivable that all the differences are encoded in a single gene, and in fact they are not. For distyly it is possible to identify at least six different genes, three affecting the female side, the style and stigma, and three controlling the anther and pollen. These genes behave as a unit in segregation, which they must do for the integrity of the system, because they are situated in a cluster on one chromosome, and probably also have abnormally low recombination by the suppression of cross-overs within the cluster. It is, in fact, the rare cross-overs that have revealed the linear order of the genes in the cluster. The genes and their phenotypes and the probable order are given in Fig. 4–1.

Short-styled						Long-styled					
G	S	I_s	I_p	P	A	g	s	i_s	i_p	p	a
g	s	i_s	i_p	p	a	g	s	i_s	i_p	p	a

Fig. 4–1 The six genes with their most probable linear order in the S gene complex of dimorphic *Primula*: G, style length; S, stigmatic papillae; I_s, incompatibility reaction of style; I_p, incompatibility reaction of pollen; P, pollen size; A, anther height. The capital letters represent the genes producing the dominant characters in the short-styled plant, and the genes with small letters produce the recessive characters of the long-styled plants.

A cross-over in a short-styled plant which occurs in the centre of the gene cluster between I_s and I_p produces a gamete with the genes, g,s,i_s,I_p,P,A; when this is combined with the genes from a long-styled plant, with which it is compatible, it produces a plant with all the stylar features of a long-styled plant and the pollen and anther phenotype of a short plant; it is a long homostyle and self-compatible similar to those in the populations of primroses in Somerset. Long homostyles are occasionally found which have high anthers, but with small pollen, and these are self-incompatible. These are the result of a cross-over between P and A, so that the cross-over gamete has g,s,i_s,i_p,p,A. The order of the last three genes is unlikely to be i_p,a,p because two coincidental cross-overs would be required to produce this self-incompatible homostyle.

The homomorphic and heteromorphic species and genera in the *Plumbaginaceae* (see Appendix) show an evolutionary sequence from monomorphic self-incompatibility, followed by pollen dimorphism, then stigma dimorphism and finally anther and style length dimorphism. Homostyle plants have also been found in trimorphic species, indicating that the S and M genes are also clusters.

In the homomorphic incompatible species, rare abnormal types cannot be found by visual inspection, but in the gametophytic system, the system itself provides a highly effective means in the style of sieving rare mutant or cross-over pollen genotypes. If we consider two species, an evening primrose from New Mexico, *Oenothera organensis*, and the European sweet-cherry, *Prunus avium*, we can see the scope and efficiency of this stylar sieve. The surface of an *Oenothera* stigma is large enough to have 5–6000 pollen grains in a single layer, the figures for the cherry are 1000–1500. All these pollen grains are in a position to germinate and penetrate the stigma; if the pollen is self pollen, then virtually all the pollen tubes will be inhibited, and no seeds will be produced. We know that there are many different S alleles, and that incompatibility is due to the presence of the same allele in pollen and style. If the S gene mutates either to another active allele or to an allele which does not function, the pollen tube carrying the mutated allele will grow down its own style and produce a seed. In some species, such as *Oenothera organensis*, with long styles and conspicuous pollen tubes, the number of mutants can be counted by counting the number of pollen tubes in the lower part of the style, Fig. 4–2, or by counting the number of seeds produced.

If the species has a large number of ovules in the fruit it is necessary to stimulate the development of the fruit with auxin because the rarity of S mutations will result in only one or two fertilized ovules in each fruit, and this is not enough, without auxin, to give the stimulus to fruit development.

In this way mutations of the incompatibility gene have been obtained in seven species; the spontaneous mutation rate varies from species to species from 0.2–10 per million pollen grains, this is well within the range of mutation rate for other genes. By growing the seeds from the mutant pollen grains, it is possible to test what kind of mutation has occurred. Most of the mutations are to self-compatibility and none to another fully-operative self-incompatible S allele. But the self-compatible mutants tell

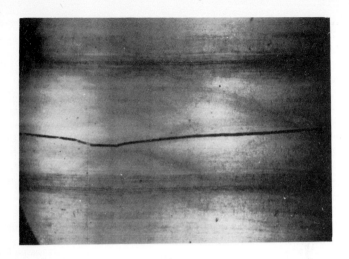

Fig. 4–2 A single mutant pollen tube in the lower part of the style of *Oenothera orgenensis* 24 hours after self-pollination with pollen derived from buds treated with 200R of X-rays. (From LEWIS, D. (1948). *Heredity*, **2**.)

us a lot about the structure of the S gene, and it has turned out to be a cluster of at least three genes. If we consider a particular S allele, say S_6 in *Oenothera organensis*, this allele has to imprint the pollen and the style with its unique specificity. A plant with S_6 and S_3 has both alleles in the style and half the pollen contains S_6 but the rare mutant pollen grain contains $S_{6'}$ which is a mutant from the S_6 allele. This pollen grain grows down the self style, gives a seed which has S_6 from the egg and $S_{6'}$ from the pollen. The plant is fully self-compatible and $S_{6'}$ pollen grows down the style; if this plant is used as pollen onto the original S_3S_6 plant it is also compatible. However, if the reverse cross is made using the S_3S_6 plant as pollen parent, then it is incompatible showing that the $S_{6'}$ mutant allele is fully operative in the style. This can be summarized in the table:

Style		Pollen	Result
S_6	×	S_6	−
$S_{6'}$	×	S_6	−
S_6	×	$S_{6'}$	+
$S_{6'}$	×	$S_{6'}$	+

Table 1 Action of mutant $S_{6'}$ allele.

The specific function of the S_6' is completely lacking in the pollen but in the style it is operating and imprints the unique specificity of S_6. Furthermore, by another generation of selfing, we can obtain S_6' S_6' homozygotes and that too rejects completely the S_6 pollen. Note that the unique specificity is unchanged and that it is fully expressed in a style which only contains S_6'; the only effect of the mutation is to prevent the expression of the specificity in the pollen. We have, therefore, separated the S complex into two genes by this mutation, one controlling S specificity, and one controlling expression of the gene in pollen. Other mutants have been obtained which have lost both pollen and stylar action.

By producing tetraploid plants which give diploid pollen grains, it has been possible to produce diploid grains with two alleles S_2S_6', and this pollen is inhibited on a style with S_2 as expected, but also on a style with S_6 or S_6'. This tells us that the missing function of S_6' in the pollen can be supplied by complementation from S_2. All the results point to an S complex of at least three genes, one gene with the S specificity and one which regulates the gene activity in pollen and the third in the style. We have no information on the linear order of these genes.

In *Oenothera*, only pollen activation mutations were found, but in *Prunus avium*, *Trifolium repens*, *T. pratense*, and *Nicotiana alata* two other types of mutation have been found. One has lost both its pollen and stylar activity and a second has lost only its stylar activity. These are of especial interest not only because they confirm the three-gene nature of the S complex, but they show that the mutations in the different genes are not entirely independent. It should be realized that the method does not select for mutations affecting the style, but nevertheless these are found. A mutation in the S specificity gene would affect both pollen and style, but a mutation affecting the style must be accompanied by some temporary reversible effect on the pollen activity. A complication has been found in S mutant in *Petunia* and *Nicotiana* because the pollen mutants are often accompanied by an extra chromosomal fragment. The best explanation of the fragment is that in these species, many of the S mutations are by themselves lethal to the pollen, and that the fragment carries a normal S gene which compensates the lethality by complementation. A discussion of these chromosomal fragments must be delayed until after the effect of polyploidy on incompatibility has been described. The mutation studies also showed quite clearly the time of action of the S complex in the development of the pollen; by irradiation to produce mutations at different stages in the development of the pollen, it was found that irradiations at, or later than, the tetrad stage of meiosis were not effective, indicating that the S complex had been expressed at this stage.

Mutation studies in the sporophytic homomorphic system do not have the same scope because of the time of gene expression and the masking effects of dominance; the highly selective stylar sieve does not select mutants. But it is highly probable that the *S* gene in the sporophytic system is also a complex of sub-genes and evidence from studies of genetic dominance in the system strongly supports such a view.

The mutations which have been observed in the gametophytic system selected by the stylar sieve are all breakdown mutations to self-compatibility either by a loss of a gene activator or by loss of the specific gene. No mutations to a new specificity have been obtained in the very extensive experiments involving at least 10^9 pollen grains. The mystery is how have the many *S* alleles which exist in the wild populations arisen. Some kind of mutation must have occured; why do we not find them in the experiments? Reports of new alleles have been made from repeated inbreeding. It may be that a new allele might arise by a series of mutations each having a small and undetectable effect. The stylar sieve would exclude these intermediate mutant stages and in this way none would come to maturity to make a fully workable new allele which would pass the self-stylar barrier. By resorting to methods and species where some self seeds can be obtained and by continued self-pollination over three generations, mutant alleles which have most of the properties of a new allele have been obtained.

4.1 Breakdown of *S* action by polyploidy

This section could have had the more dramatic title, 'The Mystery of the Giant Fertility Pear', for that is how it began. A tree of Fertility pear growing at Seabrooks Nursery sported a branch which bore thick leaves and large flowers and fruit. When examined at the John Innes Institute, England, it was found to have four sets of chromosomes instead of the normal two sets. It was tetraploid derived by a doubling up of the chromosomes. But the more surprising finding was that, unlike the diploid from which it arose, it was completely self-compatible. A straight doubling of the chromosomes had removed the incompatibility barrier. Furthermore, the pollen from the tetraploid was compatible on the style of the diploid, but the reverse cross, the pollen of the diploid on the style of the tetraploid was incompatible.

Table 2 The effect of tetraploidy in the Fertility pear.

Diploid selfed	Incompatible
Tetraploid selfed	Compatible
Diploid ♀ × Tetraploid	Compatible
Tetraploid ♀ × Diploid	Incompatible

It was clear from this that the simple doubling of the chromosomes and of the genes caused a breakdown in self-incompatibility and that the cause of the breakdown was in the pollen and not in the style.

By making tetraploids with colchicine, the same effect was found in some, but not all, *S* genotypes in *Oenothera organensis*, *Petunia hybrida*, and *Trifolium* species. In fact, in all dicotyledonous species with gametophytic incompatibility, this breakdown effect in the pollen has been found. By making a careful study of the self-pollinated styles of the tetraploid, it was found that only a proportion of the pollen grains were compatible. If we consider the consequences of tetraploidy on the *S* allele situation in pollen and style, we find the clue to the explanation. A diploid plant will have two different *S* alleles, for example S_1 and S_2. In the diploid style, both will be present; in the haploid pollen grain, only one will be present, S_1 or S_2. A tetraploid shoot of the same plant would have a style which is $S_1S_1S_2S_2$ and pollen grains which segregate S_1S_1, S_1S_2, S_2S_2 in a $1:4:1$ ratio. The new situation is not in the style which has S_1 and S_2 in both the diploid and tetraploid, but in the pollen grains which have both S_1 and S_2 only in the diploid pollen of a tetraploid. A pollen grain with different *S* alleles allows novel allele interactions. It is those heteroallelic pollen grains which are compatible and are the cause of the breakdown. In this new situation of a diploid pollen grain, the products of the two different alleles interact and aggregate to make an ineffective product. This will be explained later when the pollen product has been described.

In combinations that are not self-compatible in a tetraploid for example, S_1S_3, it is possible by suitable crosses to show an S_1S_3 diploid pollen is inhibited in a style with S_1 but not with S_3. We can say that S_1 is dominant to S_3 in a diploid pollen grain. In some combinations, for example S_3S_4, we find that the pollen is inhibited in a style with S_3 and also in a style with S_4; one can conclude that S_3 and S_4 are codominant and are both producing their effective product without any serious interaction. For the gametophytic system to work, it is essential that all alleles should be independent and show codominance in the style. If one allele were dominant over another in the style with gametophytic control of the pollen, half the pollen would be compatible on its own style thus vitiating the whole system.

It is of interest to note that this independence of *S* action is complete even in a style with four different alleles, e.g. S_1, S_2, S_3, S_4, which can easily be constructed from suitable tetraploids. Such a style inhibits pollen with S_1 or S_2 or S_3 or S_4.

Tetraploids of species with sporophytically determined incompatibility whether in a heteromorphic or homomorphic condition are fully self-incompatible and show normal diploid behaviour of their *S* alleles. This is expected when we recall that the *S* gene in the sporophytic system produces its effect in diploid cells whether it is the style or the pollen mother cell, and that the whole system depends upon dominance in the style and the diploid tissue which gives rise to the pollen.

Because the pollen is sporophytically determined, the S genes have always operated in a diploid situation where dominance and co-dominance are possible, and where alleles expressing such dominance have been selected to keep the incompatibility system efficient. Any slight complication such as tetraploidy, does not basically change this situation, and the incompatibility remains unchanged.

The two gene gametophytic system in grasses appears superficially to be an exception because although the system is gametophytic, polyploidy does not cause breakdown of S action. This will be referred to later but suffice it to point out that there has to be cooperation between the S and the Z genes at the allelic level, and because both S and Z are present in a haploid pollen grain such interactions are a normal feature. It may be that these orderly S and Z interactions are possible only when allelic interactions are not possible at the molecular level. This will be discussed after the molecular basis has been described.

5 The Molecules of Mediation

5.1 Some theoretical considerations

The knowledge of how the S genes and their alleles are expressed in pollen and style to bring about incompatibility provide a good guide to the possible molecular and biochemical basis for the reaction, and these can be accepted or rejected, in principle if not in detail, by the facts of the genetic behaviour. On the genetics alone, we can discriminate between two basically opposite types;

(1)   *A complementary stimulus* to pollen germination and growth by a complementary interaction of two *unlike* S gene products in pollen and style. This would positively stimulate pollen germination or development in a compatible combination.

(2)   *An oppositional inhibition* to pollen germination or growth by an oppositional interaction of two *like* gene products in pollen and style. This would positively inhibit pollen in incompatible combinations.

Within both of these models we can postulate that the complementary and oppositional components are either simple small molecular metabolites or large molecules with great informational content which would give great variation with high specificity. The main genetic facts to be considered are:

(i)   The number of genes – one, two, or more.
(ii)   The number of alleles – two or many.
(iii) Gametophytic or sporophytic action.
(iv) Dominance or codominance.

The absolute discrimination between the two models, complementary and oppositional, comes not from whether there are two alleles or many as one might have expected, but from the dominance or codominance of the alleles. Heteromorphic incompatibility has complete dominance of S over s in the style and pollen mother cell, this results in one phenotype 'S' in a short style and pollen, and 's' in a long style and pollen. Because the alleles are not codominant, a short style which has S and s alleles does not have both 'S' and 's' products. The genetics of this system can be satisfied equally well with either a complementary or an oppositional model. On a complementary model, 'S' would complement 's' to stimulate pollen tube growth; this would occur in the two compatible crosses Long × Short, and Short × Long, but would not occur in Long × Long or Short × Short. On an oppositional model, like would meet like, for example in the incompatible pollination Short × Short or Long × Long to cause inhibition of pollen tube growth but no such inhibition would occur in the two compatible pollinators. Both models are completely satisfied and as we shall see, there are other kinds of evidence for both.

In contrast, homomorphic incompatibility systems show codominance of the different S alleles; in the simpler gametophytic system each plant has two different S alleles and all the numerous pairs show a strict codominance in the style producing two different 'S' products. The pollen, being gametophytically controlled, receives each of these products segregated in different pollen grains in equal proportions. Any model based on complementation between unlike gene products is violated in all self- and cross-incompatible pollinations. This is shown in the typical incompatible situation, $S_xS_y \times S_xS_y$ (see Fig. 3–10).

The second model, based upon the oppositional inhibition between like allelic products, fits all self- and cross-pollinations; it also applies to one- and two-gene systems which both exhibit codominance. The same reasoning also applies to the homomorphic sporophytic system with equal force, although not all allelic pairs show codominance, the best estimate is that 60% of allelic pairs show codominance in *Brassica oleraceae*. Some 40% show dominance, but as we have previously shown, dominance and sporophytic control can be explained on either model, we are left with codominance and the exclusion of the complementary model.

The other important genetic fact to consider is the number of alleles. With two alleles as in the heteromorphic systems it is easier to conceive a complementary stimulus in that one allele supplies something that is lacking in the other; it is more difficult, but not impossible, to conceive a similar complementary system between the thousands of different pairs in a multi-allelic system. But the number of alleles is a guide to the size of the molecules interacting; with a multi-allelic system the key reaction which causes the inhibition must be between large molecules and almost certainly proteins, to give the required variation and specificity and reactivity. A two-allelic system as in heteromorphy could work with small molecules or large. We now have a theoretical framework based upon the genetics which can be tested by physiological and biochemical experiments.

5.2 Methods of study

Pollen of many plants can be easily germinated and grown in simple artificial media; the tubes grow in some species to a length nearly equal to the style through which they normally grow. The medium usually contains sucrose of varying concentrations and boric acid at 0.01%. Pollen grains are full of storage carbohydrate, sometimes starch, and the volume of pollen grains over a wide range of species and genera is closely correlated with the length of the pollen tube, i.e. the style through which it normally grows. As the pollen tube grows, the nucleus, cytoplasm and organelles and storage material are in the tip of the tube, and as the tube grows the tip is successively cut off by a callose plug. This means that the pollen grain has to make a hollow tube in some cases, as in maize or *Oenothera*, 150–200 mm long, and to convey the essential contents in the tip to the egg. Protein synthesis and some RNA synthesis occurs when pollen germinates, some organic materials used in growth also come from the style. If we could demonstrate the incompatible reaction in the artificial culture medium by adding materials from style or stigma which either inhibited or stimulated pollen-tube growth, and that this inhibition was highly specific and completely correlated with the numerous different S incompatibilities, then we would be in a position to identify directly the nature of the incompatible reaction. Unfortunately, despite many unpublished attempts by many workers, and some unconfirmed published results, no *in vitro* test has been found which is convincingly a test of incompatibility. Indirect and correlative methods of study have had to be adopted using theoretical models to narrow the search. One might look for a simple complementary system in a heteromorphic species, but not in a homomorphic system. The indirect methods are the effects of different temperatures within the physiological range to determine whether the incompatibility reaction has a Q_{10} like most chemical reactions, or the effect of heating at higher than normal temperatures to test whether the components are heat labile which is the common feature of many proteins and enzymes. Serological techniques, and chromatography and electrophoresis can be used to detect differences in protein in pollen and styles. The grafting of styles with different S alleles can be made to test whether the reactive agent in the style is readily diffusable. These and other techniques have been used to give us some picture of the biochemical nature of the reaction.

5.3   Heteromorphism

The nearest example of a complementary system is found in the distylic *Linum grandiflorum*. There are long- and short-styled plants, but the anthers are of equal height and pollen grains are of the same size. The compatible pollinations are between the two types. The clue to this system was the observation that compatible pollinated styles, and the short style when selfed, curl up, but the long type selfed does not. The style has a long stigmatic surface on one side of the style, the curling is with the stigmatic side on the concave side. This indicated that water was being withdrawn from the stigma and hence the shrinking of the cells and curling. It was also found that the long-type pollen on its own style did not swell up, but remained shrunken. The short-type pollen on its own style did swell and germinate but the tips of the pollen tubes soon began to swell and then burst.

In compatible pollination the pollen swells, germinates and grows normally. By measuring the osmotic pressure of the pollen, big differences were revealed, which are shown in Fig. 5–1.

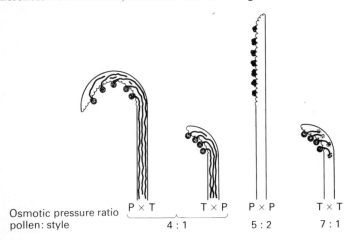

Osmotic pressure ratio P × T   T × P   P × P   T × T
pollen: style   4 : 1   5 : 2   7 : 1

Fig. 5–1 Diagram showing the relation between the osmotic pressure and the behaviour of pollen and styles. (After LEWIS, D. (1943). *Annals Bot. N. S.*, **VII**.)

The osmotic pressure in the pollen is four times greater than in its compatible style, whether the cross is Long × Short or Short × Long. In the compatible cross, or self, Long × Long, the ratio is 2:1 and the pollen does not imbibe water. In the incompatible cross, Short × Short, the ratio is 7:1, and the pollen tube bursts. The only other complementary system described was in the heterostyled garden shrub, *Forsythia intermedia,* and this was based upon two different enzymes in the styles, and two different flavenoid substrates, rutin and quercitrin in the pollen; the enzyme and its appropriate substrate were claimed to be together in the compatible combinations. Unfortunately, the observations could not be repeated; the work was a figment of a fertile imagination and is now completely discredited.

There are differences in the sculpturing of the two types of pollen and of the papillae in several heteromorphic species which are complementary, probably in helping to cause close contact between pollen and stigma, (see Fig. 3–2) but these cannot be the primary cause of incompatibility because the reaction is the stoppage of pollen-tube growth down the style.

Perhaps the best but still rather slender evidence that an oppositional inhibition is present in heteromorphic species, is the effect of temperature within the normal physiological range. *Primula obconica* is distylic; the incompatible tubes are inhibited in the style, and the inhibition is stronger, i.e. occurs after a shorter time and after less growth at 20–25°C than at 12–10°C. If we are dealing with a stimulus producing compatibility and the incompatibility is merely a lack of this stimulus, temperature should have no effect. Such a temperature effect with a Q_{10} of 2.4 is also found in several homomorphic species where an oppositional scheme based upon specific glycoproteins is known.

To summarize, the heteromorphic systems have both complementary and oppositional incompatibility with the complementary being the exception, but to have both types is in keeping with the theoretical models.

5.4 Homomorphism

The basis of the recognition between pollen and style in the homomorphic system is a protein molecule probably with glucose. This was shown by serological tests of pollen protein extracts in *Oenothera organensis* and of pollen and style protein extracts in *Petunia hybrida*; these both have gametophytic incompatibility. Pollen extracts were injected into rabbits and the serum was tested by precipitin rings against pollen protein. Specific precipitin reactions occurred only when the pollen protein carried the same S allele as the pollen injected into the rabbit. Four different alleles were used in *Oenothera* and there was complete identity. In *Petunia* it was also shown that both the S protein in the style and pollen carrying the same S allele contained the same protein as revealed by the serological test. Single pollen grains of *Oenothera organensis* could be used for the test as shown in Fig. 5–2.

By using electrophoresis to separate the proteins from pollen of plants which had been supplied with amino acids labelled with radioactive C^{14}, it could be shown that a style which had been pollinated with incompatible pollen contained a glycoprotein which carried the radioactive label and which was not present in the unpollinated styles and in compatibility pollinated styles. This evidence together with the serological experiments shows that an incompatible reaction is due to an aggregation of specific S proteins from pollen and style.

Serological tests in *Brassica oleraceae*, which has a sporophytic system, have revealed proteins in stigmas which can be related to the S alleles. However, in this plant it was not possible by these means to identify proteins in the pollen. Glycoproteins have also been implicated in *Brassica* by gel-electrophoresis. In radish, *Raphanus sativa*, another

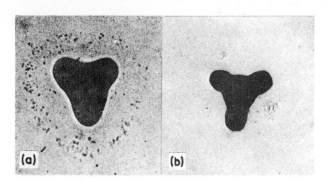

Fig. 5–2 Pollen grains of *Oenothera organensis*. (a) S_6 pollen grain. (b) S_3 pollen grain; both pollen grains are on agar with rabbit antiserum produced by the injection of S_6 pollen protein. Note the zone of precipitate in (a). (After LEWIS, 1976.)

species with sporophytic incompatibility, pollen proteins correlated with the S alleles have been found to trigger a response in stigmatic cells which is characteristic of the incompatibility reaction. The callose formed in the stigmatic cell opposite an incompatible pollen grain has been produced by the application of protein from pollen which has the same active S allele as in the stigma but not from pollen which has a different allele and is compatible. These serological and electro-phoretic tests do not prove the active molecule to be a protein or glycoprotein, but they give convincing support to what on theoretical grounds is the most likely contender.

It has been known for a long time that the inhibition in the sporophytic system is immediate and before penetration of the stigma (Fig. 5–3). It has also been known that the final inhibition of pollen-tube growth in the gametophytic system can be after several hours and after penetration of three quarters of the length of the style. But it is now realized that even in the gametophytic system, there is an almost immediate reaction in the stigma. In *Oenothera*, if the pollination is made at 30°C, the incompatible pollen tube penetrates 1–2 mm and then stops growing; this takes place in 15–30 minutes. At a lower temperature (13°C), incompatible tubes continue growing slowly for 48 hours and reach 100 mm. If the pollination is made at 30°C and after 15 minutes is cooled to 13°C, no further growth occurs. This shows that the recognition is very rapid, but that final inhibition can be delayed. It also shows that once final inhibition has occurred, it is irreversible. Further work with single and mass pollen has shown that all the S protein flows out of the pollen into isotonic media within 10 minutes, again pointing to an immediate recognition. Furthermore, if the pollen tube is examined under the electron microscope, a clear difference between compatible and incompatible

38 §5.4

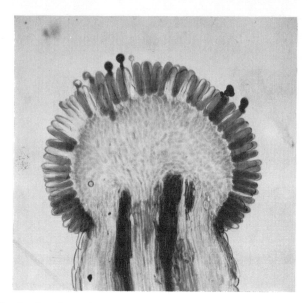

Fig. 5–3 Section through a stigma of *Raphanus sativus* after self-pollination; most of the pollen grains did not germinate, the four germinated pollen grains have not penetrated the stigmatic cell.

tubes can be seen after 15 minutes, (Fig. 5–4). By examining the changes in ribonucleic acid in the pollinated styles in *Petunia*, a difference between incompatibility and compatibility pollinated styles takes place between 0 and 3 hours, although the inhibition of tube growth is after 12 hours. We can conclude that the recognition in *Oenothera* and *Petunia* is probably by the same protein in pollen and style, and that this recognition reaction sets a change in metabolism which sooner or later stops the pollen-tube growth.

In the sporophytic system, many lines of evidence point to extreme localization of the reaction on the stigma. If the stigmatic papillae are shaved off, the incompatible pollen germinates, penetrates the style and produces seed. Figure 5–3 shows that it is germination and penetration that is inhibited. But there is a feature which may be common to both gametophytic and sporophytic incompatibility. Figures 5–5 and 5–4 show that early and excess callose is formed in incompatible tubes in gametophytic species. If we look at stigmatic cells of *Raphanus* with sporophytic incompatibility, we find a heavy deposition of callose inside the cell where an incompatible grain has become attached to the surface of that cell. Protein separated from the lipoprotein of the pollen grain when placed on a stigmatic cell, stimulates a similar callose deposition when the protein was from

Incompatible   **Compatible**

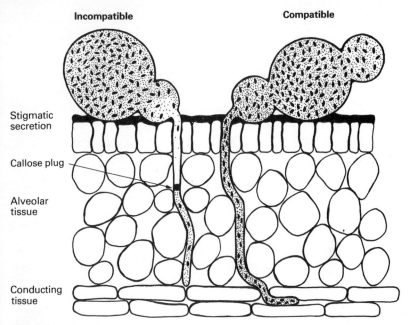

Stigmatic
secretion

Callose plug

Alveolar
tissue

Conducting
tissue

Fig. 5–4 Diagrams of incompatible and compatible pollen tubes 30 minutes after deposition on the stigma in *Oenothera organensis*. Note the callose plug in the incompatible tube. The frequency of the small black dots represent the concentration of free carbohydrate and the black rods denote the starch grains.

incompatible pollen. This is good evidence for the protein nature of the incompatibility molecule.

The action, whether it be sporophytic or gametophytic, is highly localized in the sense that pollen tubes side by side go their own way, one is inhibited, the other continues unaffected. Style-grafting experiments have been done to test whether there are diffusable substances in the style which cause incompatibility. The evidence from a number of species is somewhat obscure, but the most convincing is that no diffusion of the actual reacting molecule occurs.

Many different treatments of the styles to test for attenuation of incompatibility have been made. These have ranged from hot-water treatment, CO_2, to massive doses of X-rays, and in several different species. These have in general weakened the incompatibility reaction. This is important because this would not be expected if the action is a complementary stimulus, but would be expected on the oppositional system.

There is one final point of importance, and that is the place of origin of the *S* protein for the pollen. The evidence in *Raphanus* is

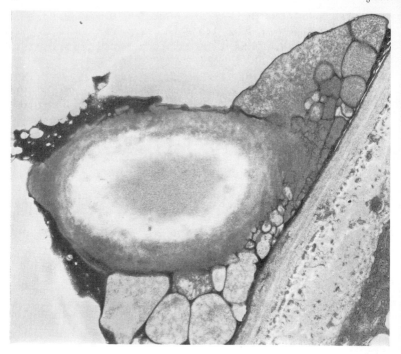

Fig. 5–5 Section through grain (on left) and the surface layers of the stigmatic cell (on right) of *Raphanus sativus*. The pollen and stigma are incompatible. Note the tube has not penetrated the stigmatic surface, and that a layer of white callose has formed at the place of contact. (Figures 5–4 and 5–5 by courtesy of Dr Hugh Dickinson.)

that it is synthesized in the nutritive tapetum tissue, and that it is deposited on to the pollen grains before leaving the anther. This is possible and would be in agreement with sporophytic control. Such an origin for gametophytic incompatibility is not possible, because the synthesis must occur after the segregation of the alleles at meiosis in the pollen mother cells. Mutation of the *S* gene by X-rays at different stages in meiosis has shown that the *S* gene has produced its primary product, messenger RNA, or secondary product, protein, at the time of the tetrad stage.

To end this molecular aspect of incompatibility, some speculations based upon the genetic facts and what we know about proteins as enzymes or structural elements are given. It may be wrong, but it may help to guide future investigations. With any molecular basis, we must be able to explain in the gametophytic system the strict co-dominance of *S* alleles in the style, and the dominance and negative complementation which is shown by the same allelic pairs in artificially produced diploid pollen grains. Natural proteins and enzymes are polypeptides of some hundreds of amino acids. Some active proteins in the cell are single polypeptides, they are monomers, others exist as multimers with 2, 3, 4...12 identical polypeptides aggregated into a functional whole. Still others are aggregates of two different polypeptides, the mammalian haemoglobin is a good example of a tetramer with two α chains and two β chains. It is known that some dominance, and all allelic complementation and negative complementation, are properties of hybrid multimers with polypeptides from different alleles.

The fact that codominance is the rule in the style suggests that the active protein is a monomeric protein, e.g. in the $S_1 S_2$ style, S_1 gives α_1 and S_2 gives α_2 as separate allelic products. In the haploid pollen grain, there is only one S allele, and this could give $\alpha_1 \alpha_1$ dimers or $\alpha_1 \alpha_1 \alpha_1 \alpha_1$ tetramers without any chance of hybrid multimers such as $\alpha_1 \alpha_2$ being produced. There is, therefore, no restriction in the pollen from the system which requires codominance in pollen and style to develop a multimeric protein. The fact that these dominance interactions are found in diploid pollen grains where hybrid multimers would be formed gives very powerful support to this hypothesis. The genetics therefore indicates that the recognition molecules are monomers in the style, but multimers in the pollen. The biological advantage of this system is that the advantage of multimers can be exploited without the restriction of unwanted dominance interaction.

If we carry the hypothesis to the grasses with S and Z genes, then the haploid pollen could be restricted to an SZ dimer; this gives the unique complementary specificity and the lack of dominance interactions between the alleles.

The sporophytic system has codominance, dominance and negative complementation in both pollen and style. Both pollen and style are similar in that the relevant gene action occurs in diploid tissue in both. Strict codominance might have evolved on a monomeric protein, but this is unlikely because for a sporophytic system with codominance to evolve, it must have at least four different alleles at the very start, and even then, only one in five of the pollinations would be compatible. This appears to be too big a price to pay for strict codominance with sporophytic control. This leaves a multimeric protein to give the dominance and it must be multimeric in both pollen and style. The lack of breakdown by polyploidy in the sporophytic system supports this.

From these theoretical speculations and the experimental facts it is clear that study of the molecular basis is in an active state, for our ignorance far exceeds our knowledge.

6 Incompatibility and Man

I have been discussing incompatibility as an evolutionary development in plants to ensure their future fitness, and as if it were of interest only to those who are curious about nature and particularly about the nature of plants. But incompatibility has an important impact in our methods of cultivating and breeding of plants; and it almost certainly plays a major part in causing hay fever.

Self-incompatibility occurs in twenty or more cultivated plants; whether it is a nuisance or a benefit to the grower or plant breeder depends upon how the crop is grown and propagated, and what part of the plant is harvested. Fruit such as apples and cherries which are dependent on seed formation, or a seed crop such as pea or wheat, all depend upon effective fertilization. In leaf crops such as cabbages or sprouts the grower is not interested in seed formation. This is the concern of the plant breeder and seedsman; the grower wants uniform plants which will give him a high yield of uniform sprouts of the right type: the ideal for the seedsman is seed that he can produce easily and if possible keep under his own control. In these two extremes of plants, the fruit and the cabbage, we will see how self-incompatibility has been a nuisance and also a benefit, and how geneticists and plant breeders have been able to overcome the nuisance and utilize the benefit.

6.1 More reliable fruits

It was a change in the methods and economics of the cultivation of temperate fruit trees which gave self-incompatibility an agronomic importance beyond that recognized by biologists. As a rarity, old farm orchards may still be seen today, in which there are almost as many varieties as trees. Many of the trees were directly descended from a stone or pip ensuring that each would be genetically different and therefore cross-compatible. The very names, Costard, Forge, Curltail, had an old-fashioned sound. The beginning of modern marketing and transport stopped all this; a uniform fruit was required in large quantities for packing and transporting to large cities. In the middle of the nineteenth century, farm orchards were replaced by seried ranks in large acreages of a few, well-tried varieties. This was when the pollination trouble began. Large areas of one variety would not set adequate fruit because all the trees were genetically identical and had the same S alleles. They were not only self-, but between trees, cross-incompatible. To overcome this, two varieties were interplanted together. This normally worked but mistakes were made because although the two varieties might be quite different in their habit and fruit, they had the same two S alleles and were cross-incompatible.

By studying the S alleles in varieties of apples, pears, plums and cherries, compatibility rules were drawn up so that pairs of different fruits could be selected to give mutual cross-compatibility. Careful testing of different varieties revealed that they could be classified into groups, all crosses within a group were incompatible but between groups all were compatible. It was also possible by testing the next generation from such crosses to show that the temperate fruit trees of the *Rosaceae* have the homomorphic gametophytic system.

It was also found at the same time that some fruits had three sets of chromosomes and produced sterile pollen; they were not suitable as pollinators for any variety, not because of incompatibility but because of pollen sterility. Blenheim Orange and Bramley's Seedling are two such triploid apples.

There are other practical restrictions on the selection of interfertile pairs of varieties; they not only have to be compatible, but they must flower at the same time. The problems did not end here because the conservative thrifty bee concentrates its efforts on one tree unless forced to move on by pressure of numbers, and is a very good self-pollinator. There must be a high, active bee population to ensure cross-pollination and a crop. This requires many bees and good weather. The massed orchard means fewer natural habitats for wild bees, so that hives of domestic bees are installed, but the weather can still defeat the whole system. Also a further disadvantage to the grower is the complication of spraying and picking a mixed orchard. Much of this can be overcome by having self-compatible varieties when these do exist, in apples and plums. The most regular cropping and most northerly grown are the self-compatible varieties; Victoria plum is a good example. Peaches present no problem because they are all self-compatible, so are Morello cherries, but the sweet cherry is strictly self-incompatible in all wild and cultivated varieties.

The stylar sieve of self-pollination described earlier was used thirty years ago on a big scale in sweet cherries with and without mutagenic treatment to produce fully self-compatible seedlings. With these seedlings, there was no problem in producing a whole range of new, self-compatible sweet cherrries. Such a seedling from the variety Bigareau Napoleon, which is $S_3 S_5$, has S_3 and S_c alleles, the S_c being the new mutant allele. When this is used as a pollen parent on a whole series of good varieties which have $S_3 S_5$, all the progeny carry S_c because only S_c pollen is compatible, and all the progeny are self-compatible. Where are these self-compatible cherries today? Except for a variety in Canada they are either destroyed or still on trial. The official reason for their absence is that commercial cherry growing in Great Britain is not an economic proposition. This is not surprising if similar developments are treated in the same way, but in any case the private grower might have enjoyed first the blossom and then the fruits, if protected from birds, of a single cherry tree.

6.2   Better Brussels sprouts

Self-incompatibility has been utilized successfully in breeding better cabbages, Brussels sprouts and fodder kale by the production of F_1 hybrids. The great advantage of F_1 hybrid crops was realized and commercialized fifty years ago in maize, *Zea mays*. The advantage was that F_1 hybrids between good combining parents had a higher yield and were more uniform and adaptable than non-hybrid varieties. The fact that it became a commercial possibility in maize was two-fold; firstly the plant is a natural outbreeder due to the temporal difference in maturity of pollen and stigmas, and the advantages of F_1 hybrids are more pronounced in an outbred species. Secondly, the anthers, borne on 'tassels', at the apex of the plant could easily be removed by machine without touching the female cobs which were three feet below in the axil of a leaf. Maize is not self-incompatible, and self-pollination can be used to obtain inbred lines. When a pair of inbred lines that combine on crossing to give a good F_1, the F_1 seed can be made in commercial quantities merely by planting alternate rows of the two inbred lines, and cutting off the anthers of one line. All seeds produced on this line will be F_1 hybrids. More recently, the mechanical castration has been replaced by the use of a genetic factor which causes male sterility.

In a crop plant which has its anthers and styles in the same flower, such as the cabbage, it is not feasible to emasculate on a large scale to produce a commercial crop of F_1 hybrid seed. A genetic means of controlling the seed production must be used. The new hybrid cabbages, Brussels sprouts and kales owe their origin to sporophytic incompatibility. This system, as pointed out earlier, has not only sporophytic control of the pollen, but dominance and recessiveness of alleles, and because of this, homozygotes of recessive alleles are readily produced. For example, if S_1 is dominant to S_2, then by crossing $S_2 S_3 \times S_1 S_2$, one quarter of the progeny are homozygous $S_{2,2}$, by obtaining two different homozygotes they can be intercrossed to give all F_1 hybrids.

$$\textbf{Parents}\quad S_{2,2}\ \times\ S_{4,4}$$
$$\downarrow$$
$$\textbf{F}_1\ \textbf{hybrid}\quad S_{2,4'}$$

Other F_1 hybrids can be produced by having two further homozygous crosses, for example $S_{5,5}\ S_{6,6}$. By combining these, an $F_1S_{5,6}$ hybrid is produced. These two F_1 hybrids can be obtained in large amounts and then interplanted on a large scale, the result is a double-cross hybrid.

$$
\begin{array}{cc}
F_1 & F_1 \\
S_{5,6} \times & S_{2,4} \\
& \downarrow \\
S_{5,2}\ S_{5,4} & S_{6,2}\ S_{6,4}
\end{array}
$$

This method works because the only compatible pollinations are between $F_1\ S_{5,6}$ and $F_1\ S_{2,4}$ so that all seed is of a double-cross hybrid nature. To obtain good, commercially acceptable hybrids, the homozygotes must be bred with many different genetic backgrounds and tested in several combinations before a good hybrid is produced. Once this has been attained, the good combining lines can be propagated by vegetative division or by selfing in the bud stage before the self-incompatibility has developed. This 'bud fertility' can also be used in the production of the original homozygotes.

The double-cross hybrid has the advantages of facilitating the introduction of specific genes for quality, yield or disease resistance, it also gives a more economical seed production because the F_1 hybrids are more vigorous than the homozygous inbred lines used in the production of the F_1. This successful application was only possible because of the complete understanding of the sporophytic system and the careful work to test some forty different alleles for their dominance and codominance.

The contrast between the fruit tree – which is propagated by vegetative grafting where all individuals of the same variety are a clone, the crop is the fruit and the self-incompatibility is a nuisance – and the Brussels sprout – propagated by seed, the crop is the vegetative part and the self-incompatibility can be a useful breeding tool – emphasizes the diversity of situations arising in the cultivation and breeding of plants. In ornamental plants propagated by seed, chance colour or other mutations are often a source of new varieties. The plant breeder isolates the single mutant to obtain a pure-breeding line. If the species is self-compatible, then what seed is produced is from the stray pollen which gets through the isolation, and the mutant, usually recessive, is lost for a generation. Only by breeding a further generation can two plants with the same mutant colour be crossed compatibly to produce a line pure-breeding for the colour. Attempts many years ago to get a white *Linum grandiflorum* by a famous seed firm were baulked because the self-incompatibility of the species was not recognized.

F_1 hybrids are also of value in ornamental flowering plants, because when they have been produced from S homozygotes, they are all cross-incompatible, and they produce few seeds; this prolongs flowering and pleases the grower, and their lack of seeds pleases the seedsman.

6.3   Improved cacoa

Cacoa is an anomaly in the incompatibility world (see § 3.6), and is a crop of great economic importance. Like the cherry tree, the crop is in the fruit and depends upon compatible fertilization. It also, like many other plants, suffers from virus diseases which are transmitted vegetatively in the propagation by cuttings or grafts. Also, the high quality cacoa, from which good chocolate is made, comes from self-incompatible stocks. Instead of vegetative propagation, a method of producing F_1 hybrids from seed is rapidly replacing the old method. Parental lines are increased vegetatively under conditions as free from virus as possible. Parental clones are planted together, and the seed produced is all F_1 seed because the clones are self-incompatible. Unlike the cabbage, where the commercial crop is not dependent upon fertilization, the cacoa parental lines must not be S homozygotes, otherwise the F_1 would not produce a crop of cacoa pods. With both parental lines being S heterozygotes, the F_1 progeny will be similar to a four-group family (see Fig. 3–8), and there will be enough sib cross-compatibility to produce a crop. This should improve the reliability and the quality of the cacoa by reducing the frequency of virus transmission, and the quality will also be improved by the greater opportunities for breeding in new genes; but all this will probably have no effect on the fluctuations of world cacoa prices.

6.4   Hay fever

Self-incompatibility may also be important in the allergy caused by pollen, hay fever. There is not a complete correlation between the self-incompatible species of pollen and their ability to act as allergens, but certainly many of the species, particularly the grasses, which have notorious allergens, are self-incompatible. It is also true, as will be explained later, that many species that are self-compatible today have arisen secondarily from self-incompatible ancestors. (The evidence for a connection between self-incompatibility and hay fever is that the S protein flows out of pollen within minutes of touching an isotonic medium, so does the allergenic protein. In some species tested the S protein can act as an antigen when injected into rabbits.) If this connection is real and the S protein is the main allergenic protein, then the spectrum of different specificities from the very large number of S alleles even within a single species, complicates beyond the present conception, the clinical testing and desensitization of sufferers from asthma and hay fever.

7 Fungi, Slime Moulds and Other Lower Organisms

Flowering plants have evolved incompatibilities of great versatility and efficiency to suit a diploid hermaphrodite. Other vascular plants such as conifers and ferns have also adopted incompatibility as an outbreeding method, but too little is known about them to be able to make profitable comparisons. Bacteria, fungi, algae and slime moulds have incompatibility systems which operate between confronting individuals that are haploid and about which sufficient is known to be able to examine not only the types of systems within this group, but profitably to compare them with those in the flowering plants. Where there is sufficient knowledge about both sides, the comparison can be illuminating by suggesting a further, as yet undiscovered system, which had not been realized without the cross-fertilizing of ideas obtained from fungi and plants.

Bacteria are prokaryotes with their genome a naked closed loop of DNA instead of the chromosomes with histones and DNA contained in a nucleus. They have their own breeding systems which are two inter-fertile breeding types, F^- and F^+. The fertility factor in F^+ is found in several different forms, plasmids, and these can carry small amounts of genetic material and can be incorporated and recombined in the bacterial DNA. These plasmids can be classified into groups such that plasmids from the same group cannot replicate together in the same cell, whereas plasmids from different groups can. Plasmids and the F^-/F^+ system have given bacteria great scope for maintaining and recombining their genomes. These bacterial systems are so different from those in plants that no useful comparison can be made with them, but the F^-/F^+ and the plasmids exploit the peculiar naked DNA condition of bacteria. The evolutionary success of these systems has recently been adequately demonstrated by the rapid acquisition by bacteria of multiple resistance to antibiotics, and it is included here in order to emphasize the essential importance of outbreeding and recombination in all organisms from bacteria to man.

The most useful comparison is with the fungi, but before describing these, a brief mention of the slime moulds and ciliates is necessary for completion. Both of these organisms have developed breeding systems based upon multiple alleles, six or more in slime moulds, and in this respect are the only known groups outside the flowering plants and fungi that have developed this efficient type of outbreeding. As in flowering plants, matings between any two different mating types are compatible, but pairs of individuals from any one mating type are incompatible.

The most useful comparison comes from fungi. The mating components in most fungi are haploid and they have no diploid structure analagous to the style in plants. It is this difference in life form that makes the comparison so informative.

Unicellular yeast, *Saccharomyces cereviseae*, probably the simplest fungus, exists in two mating types, *a* and α. A compatible mating is between single cells with different mating types, and a compatible mating results in clumping and cell fusion followed by nuclear fusion to form a diploid cell which then can divide by budding as a diploid clone, or undergo meiosis and produce four haploid spores. The four spores are analagous to the gametes in plants and represent the products of the same individual, the mating of any two products of the four is equivalent to selfing in a diploid organism. In the tetrad, two spores will carry mating type *a*, and two, mating type α. Of the four possible pairs of matings two will be between *a* and α, one between *a* and *a*, and one between α and α; only half the matings are incompatible. So unlike the flowering plants which even in the two mating type system of heteromorphy give 100% protection against selfing, protection against selfing in yeast is only 50%. This reflects the difference between a haploid × haploid situation in yeast, and the diploid × haploid situation in plants. The ascomycetous filamentous fungi exemplified by the bread mould, *Neurospora crassa,* have virtually the same system as yeast. It is perhaps strange that these fungi have not found more efficient methods by adopting a multiallelic system; only in the higher fungi, the Basidiomycetes have multiallelic and digenic control evolved to improve the system. The yeasts and *Neurospora* are very interesting from the molecular point of view; the two mating types differ in a glycoprotein which gives complementary binding properties to cells of different mating type.

The Basidiomycetes, as exemplified by the ink-caps, *Coprinus comatus, C. cinereus* and the wood rotting fungus, *Schizophyllum commune,* and many others, have evolved two types of incompatibility. Firstly, there is the *bipolar*, as found in *Coprinus comatus,* which is controlled by one gene, *A*, with multiple alleles. From any one mating, e.g. $A_1 \times A_2$, the basidiospores segregate into two different mating types, A_1 and A_2, only matings between the different alleles are compatible. This, as in yeast gives only 50% direct protection against selfing. However, the large number of alleles in the population as a whole means that an outcross to a spore from another fruiting body will approach 100% compatibility in direct relation to the number of alleles in the population. The actual selfing in a population will therefore be greatly reduced. The second type, *tetrapolar*, found in *Coprinus cinereus* and *Schizophyllum commune,* lead to an even greater reduction of selfing by the addition of another independent gene, B, also with multiple alleles. Compatibility is obtained only when the two mating individuals differ both in their A and B alleles. For example, two haploid colonies, A_1B_1 and A_2B_2 crossed together are compatible, and give a fruiting body, (mushroom).

Parent
colonies $A_1B_1 \times A_2B_2$
              $\downarrow$

Spores
progeny A_1B_1 A_1B_2 A_2B_1 A_2B_2

The spores from the fruiting body give four different mating types, and when these are intercrossed, only 25% of the crosses are compatible as shown in the matrix given below:

	A_1B_1	A_1B_2	A_2B_1	A_2B_2
A_1B_1	−	−	−	+
A_1B_2	−	−	+	−
A_2B_1	−	+	−	−
A_2B_2	+	−	−	−

Because intercrossing haploid spores from the same fruiting body is genetically equivalent to selfing, this system gives direct protection against selfing of 75%. The presence of multiple alleles at both genes means that actual selfing in a natural population would be reduced to negligible proportions.

The higher fungi, without the advantage of diploid tissue, have attained nearly the same efficiency as higher plants by resorting to two incompatibility genes. These two genes operate in a very different way from the two genes in grasses. In the grasses, the two genes, S and Z, complement to give *incompatibility*, both S and Z alleles must be the same to give incompatibility, a system which gives none, or but a little advantage over a one-gene system.

In the Basidiomycetes, the two genes, A and B, complement to give *compatibility*, and for this there must be different alleles of both A and B. The lack of allelic difference at either A or B leads to incompatibility.

This is a fundamental difference which is a consequence of the haploid/haploid situation and the usual life cycle of the fungus. The basidiospore colonies are haploid monokaryons with one nucleus in

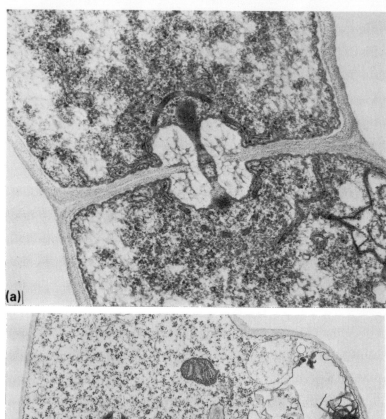

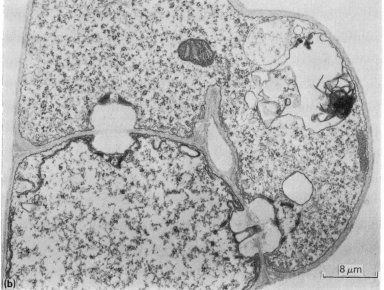

Fig. 7–1 Electron micrograph of (**a**) cross wall with pore and guards (dolipore), and (**b**) clamp connection in *Coprinus lagopus*. (Courtesy of Dr L. Casselton.)

each cell; a compatible mating produces a stable dikaryon with two nuclei per cell, which can persist for a very long time and produce fruiting bodies in which nuclear fusion occurs, followed by meiosis and the production of haploid basidiospores.

The dikaryon is a unique type of structure which maintains itself by a more or less synchronous division of the nuclei and the formation of a clamp connection, a unique structure in all nature. Another feature of importance is the complex nature of the cross-partitions in the hyphae (Fig. 7–1). These have a central pore, not large enough to allow a nucleus to squeeze through, surrounded by two hemispheres with smaller pores. The meaning of these odd features was a mystery until illuminated by the action of the incompatibility genes. The A gene is necessary for the formation of the back-growing hyphae which form the clamps; this was found by mating monokaryons which have common A alleles but different B alleles, e.g. $A_1B_1 \times A_1B_2$. This mating shows some of the features of the formation of a dikaryon but it does not produce the clamp hyphae. But why have the clamps? The answer is revealed by the action of the B gene. One of the steps in the compatible formation of the dikaryon is the migration of the nucleus from one monokaryon into the mycelium of the other. To do this, the complex dolipore septum must be dissolved so that the nucleus can pass from cell to cell. It is the B gene which controls this dissolution by the production of an enzyme, R glucanase; the enzyme is produced when there are different B alleles in the mating monokaryons. If we look more closely at the position of the nuclei in the cells after hyphal fusion, we can find a very interesting regularity which contrasts with the nuclear position in flowering plants. In the fungus, the invading nucleus always passes from a dikaryotic cell to a monokaryotic cell, whereas in the plant, the invading pollen nucleus is always from a haploid monokaryon to a diploid. This sets the pattern for the possible action of the incompatibility genes. In the plant for this reason, the action of the S alleles is one of oppositional inhibitor between the same alleles to give incompatibility; in the fungus, the action is a complementary stimulus between different alleles to give compatibility (Fig. 7–2). This has been verified by making diploid monokaryons with two different B alleles, B_x and B_y. These form normal dikaryons with either B_x or B_y monokaryons. To turn to the clamp connection, the reason for the clamp is now clear because the B gene also controls the dissolution of the clamp tip and the point of contact with the cell wall of the penultimate cell. A nucleus with a different B allele migrates from the clamp of the dikaryotic tip cell to the monokaryotic penultimate cell. This is very similar to the migration of the nucleus through a monokaryotic mycelium. This gives a unity to the B gene action and the reason for the unique clamp. It also highlights the types of gene action, whether complementary or oppositional, that can work in plants and fungi. We shall see in the next chapter that a new kind of

52

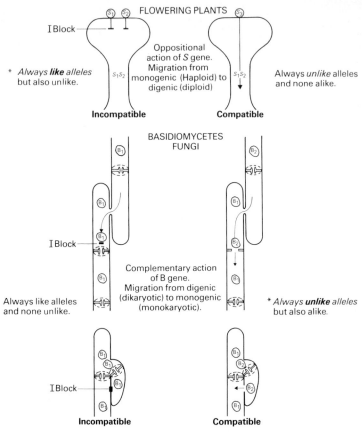

* The only situation unique to one type of mating and therefore the operative situation.

Fig. 7–2 Diagram to illustrate the action of the *B* gene in Basidiomycetes and the *S* gene in plants.

cyclical incompatibility is a possibility in plants by adapting a complementary system from the fungi.

There is a further complication in the Basidiomycetes of great interest in relation to the two genes *S* and *Z* in grasses, and the complex nature of *S* gene in the gametophytic system. The mating type genes *A* and probably *B* are each really two separate components called α and β. These are linked and they cooperate to give a unique specificity like the *S* and *Z* genes in grasses.

8 Evolution of Incompatibility: the price of a gene bank and long-term insurance

The basic assumption throughout this survey of incompatibility is that outbreeding is a necessary adjunct to sexual reproduction for the maintenance of a gene bank for genetic diversity. The sudden development of the flowering plants in the Cretaceous period has, by some authors, been attributed to the closure of the ovary and the development of the style, which made the highly efficient incompatibility system a possibility. It can also be argued that a general species recognition between pollen and stigma and self-incompatibility evolved side by side to suit the somewhat hazardous pollination by wind and insects. What is more difficult to visualize and estimate is the immediate premium which would have to be paid by individuals or small populations to obtain the insurance against future eventualities. The premium would be in fertility currency, which for those individuals that are self-incompatible and in a minority, would suffer a reduction. It is possible that self-incompatibility started as a less rigorous system in which fertility from selfing was only partially reduced and as the system spread in a population, the self-incompatibility would become stronger. Species with weak self-incompatibility are known today and it has also been possible to select for weak self-incompatibility in several species. We can also get away from the dilemma by assuming that in a critical period of rapid expansion, as occurred in the flowering plants, the immediate exploitation of many micro-environments was a great advantage, and therefore a return on the premium would be immediate.

The one safe conclusion that can be drawn about the evolution of self-incompatibility is that it is ancient, and must have been established early in the development and diversification of the flowering plants. Certainly, self-incompatibility predates the diversification into families, because families of plants have their own characteristic system. The type of system is also correlated with a basic difference in the development of the pollen grains which are characteristic of taxonomic groups. Mature pollen grains may be either binucleate or trinucleate, the binucleate pollen undergoes a division of the generative nucleus in the pollen tube. Species with sporophytic homomorphic incompatibility have trinucleate pollen grains while those with gametophytic incompatibility have binucleate pollen grains. The heteromorphic systems are found in both trinucleate and binucleate species. A second distinction of pollen is the type of division of the pollen mother cell, whether simultaneous or successive; this is also a taxonomic character and is also correlated with the system of incompatibility.

A fact of major evolutionary importance is the present-day occurrence of thriving inbreeding species which are self-compatible. The common groundsel, *Senecio vulgaris*, is a good example. Careful study of these inbred species has shown that they have been derived from outbreeding self-incompatible species. They have all the floral features of their outbred ancestors to attract insects, but these are reduced to insignificance. The four species of passion flower, *Passiflora*, illustrated in Fig. 8–1 are a striking and typical example. In the small dull green

Fig. 8–1 Flowers of three large-flowered species of passion flower and one small-flowered species, *Passiflora rutilans*. The large-flowered species are self-incompatible and the small-flowered is self-compatible. (From LEWIS, D. in *Reproductive Biology and Taxonomy of Vascular Plants*. (Conf. Bot. Soc. of the B.I.) (1966). Ed. J. G. Hawkes. Pergamon Press, Oxford. 183 pp.)

flowers of *P. rutilans* all the features of the larger highly coloured flowers are present, but reduced. The self-compatible *P. rutilans* has evolved from a highly decorative, insect-attractive and self-incompatible ancestor.

We have seen in a previous chapter how a change by a single mutation can bring about self-compatibility. The lack of an insect pollinator in an isolated situation could mean extinction to a self-incompatible species, but the isolated self-compatible mutant could survive, and be the progenitor of an inbred population. One common feature about self-compatible species in an otherwise self-incompatible genera is that they are often annuals and have a restricted habitat.

There is very good evidence that in heterostyled plants the unusual homostyle self-compatible populations are derived and not primitive. Tristylic species are confined to families which have two developmentally distinct whorls of anthers. Some species in an otherwise tristylic family are distylic, again the evidence is that these have evolved from a tristylic ancestor by loss of the mid-styled form. This has been found in two of the three families which are tristylic, the *Lythraceae* and *Oxalidaceae*. All this is evidence for the ancient origin of all incompatibility systems, and that today's inbreeders are derived from them.

A relationship between closely related self-incompatible and self-compatible species has been found by testing the compatibility between them; and a general rule, with some exceptions, is that when the self-incompatible species are used as male, the cross is compatible, but the reverse cross is incompatible. It has been termed *unilateral interspecific incompatibility*. From a full analysis of the situation, there is evidence that a mutated S gene is still operating and takes part, but not the sole part, in causing this unilateral incompatibility. One feature of these interspecific incompatibilities is that the unilateral effect is found even between two related families, the *Scrophulariaceae* and the *Solanaceae*, so the gametophytic incompatibility must have predated the diversification from a common ancestor.

Perhaps the most striking and intriguing finding has been the recent discovery of incompatibility controlled by three and four separate genes, each with multiple alleles and each complementing to give incompatibility. This has been found in *Ranunculus* and *Beta* with a gametophytic system, and in *Eruca sativa* with a sporophytic system. It is possible to argue that this multigenic system is primitive, and the common one-gene system is derived by progressive homozygosis of all but one of the genes. If this is so, it will completely change our ideas about the molecular and evolutionary basis.

I have pointed out that the efficient use of pollen and the control of sib-mating are important features in plant incompatibility systems. If we compare the different systems for these features as in Table 3, we see the severe limitation of heteromorphy for both features. The one-gene homomorphic systems are much more efficient. The multigene systems are very inefficient at controlling sib-mating, but slightly more efficient in the total use of pollen. The two features here are opposing and are presumably balanced by natural selection. Which has been most important is impossible to say with our limited knowledge of the amount of sib-pollination that occurs in natural populations. On balance, the evidence indicates that the multigenic systems were primitive, and the common one-gene systems are derived.

We have seen how the special needs of the fungi have been met in their own different way; and now, with the multigenic systems established, it is not unprofitable to look for possible incompatibility systems in plants which are derived from some features in the fungi. If we adopt the complementary B gene actions of the Basidiomycetes, which gives compatibility between unlike alleles, and multiple alleles, and assume the

Table 3   A comparison of the efficiency as estimated by the percentage of sib-compatibility and of effective pollinations.

	Genes	%effective pollinations Pop.	Sib	Sib/general compatibility	With 50% sib pollination %of effective pollen
Heteromorphic					
Dimorphism	1	50	50	1.00	50
Trimorphism	2	33	28	0.84	31
Homomorphic					
Gametophytic:	1	100	47	0.42	73
duplicate	2	>100	6	0.06	53
complementary	2	100	72	0.72	86
complementary	4	100	92	0.92	96
Sporophytic:	1	100	25–50	0.25–0.5	62–75
Sporophytic:	3	100	75–97	0.75–0.97	87–98

genes, A, B, and C, we can construct a system in plants which produces incompatibility alternately between generations. If we assume a plant to be $A_x A_y$ in which compatibility is obtained by unlike alleles, then this heterozygous plant is self-compatible. A_x pollen can be complemented by A_y in the style and *vice versa*. The progeny from such a selfing would be $1:A_x A_y$, $2:A_x A_y$ and $1:A_y A_y$, so that half the progeny, the homozygotes $A_x A_x$, and $A_y A_y$, will be self-incompatible, but cross-compatible. If the number of genes is increased to three and each gene acts in a complementary way between alleles, a plant which is heterozygous for all three genes will also be self-compatible, $A_x A_y$, $B_x B_y$, $C_x C_y$. When such a plant is selfed, the progeny will consist of one-eighth self-compatible which are heterozygous at all these genes, and seven-eighths self-incompatible, which are homozygous at any one or more of the three genes. From any compatible cross between the self-incompatible sibs, the next generation will consist of one quarter, one half and all self-compatible plants according to the combination. If the self-compatible plants are cross-pollinated from plants of other families, with multiple alleles the proportion of self-compatible will be about 100%. We have here a system of alternating self-incompatibility with self-compatibility which gives a delicately balanced breeding system; if there is an excess of selfing in one generation, this will be balanced by more crossing in the next. Perhaps such a system will be found because it is inconceivable that evolution has not found it.

Wherever we look in the living world, be it a replicating DNA in a virus or even in the test tube or a bacteria, plant or man, there is the facility and encouragement for recombination of genetic material, and this appears to be as important as the replication of the DNA itself.

9 Practical Work with Pollen, Styles and Incompatibility

9.1 Staining for pollen tubes in the style

9.1.1 Radish and cabbage

There is a simple test for incompatibility in radish and cabbage in the *Cruciferae* and in *Cosmos bipinnatus* in the Compositae by staining the pollinated style 24 hours after pollination. The style is fixed for at least 30 minutes in 3 parts absolute alcohol to one part of glacial acetic acid, although alcohol alone will do. The style is then placed on a microscope slide, gently squashed and 2 or 3 drops of the stain (given below) is placed on the squashed style. This is left to stain for about an hour before applying the cover glass.

Stain for pollinated styles

Acid fucsin 1% aqueous	2 ml
Light green 1% aqueous	2 ml
Lactic acid	40 ml
Water	46 ml

The incompatible pollen will be full of cytoplasm and stained pink and either not germinated or with a small swollen tube which has not penetrated the stigmatic cell. Compatible pollen will be empty with an empty tube penetrating into the stigmatic cell and style. Lower in the style, pink stained pollen tubes may be seen in good preparations.

9.1.2 OENOTHERA *species, evening primrose*

The pollen of *Oenothera* species contain large oval starch grains. These pass into the growing pollen tube. Styles can be fixed and stained in a solution of iodine in potassium iodide and alcohol. (Iodine 1 g, KI 1.5 g, 30% alcohol 100 ml). The styles can be examined within half an hour by simply squashing the styles in the stain on a microscope slide. Note that European wild *Oenotheras* are not self-incompatible but species such as *O. organensis* and *O. missouriensis* from North America are. The willow herbs, *Epilobium* species, also have conspicuous starch which pass into the tube.

To test whether a plant is self-incompatible it is necessary to protect the flowers from stray unwanted pollen whether by insects or wind. This can be done by placing the plant in an insect-proof greenhouse or cage or if the plant is very large, such as a fruit tree, to place a waterproof but not air-tight bag over a cluster of flowers. The appropriate pollination is then done with a fine brush or by applying a dehisced anther with a pair of forceps. The pollinated flower can either be fixed 24 hours later for observation of pollen germination and pollen tube growth or left to observe whether the fruit and seeds are produced.

9.1.3   Fluorescent staining for callose plugs in pollen tubes

This method requires a microscope with direct illumination from an ultraviolet lamp with an emission wavelength of approximately $356 m\mu$, suitable filters and a dark room for observation. The pollinated styles are fixed in formalin, acetic acid 80%, alcohol in a 1:1:8 ratio for 24 hours, cleared to make the material translucent in a $\pm$ 8 M NaOH solution for 24 hours or more until clear. The prepared styles are stained in water soluble aniline blue dissolved in 0.1 M $K_3 PO_4$. The styles are mounted in a few drops of the aniline blue stain. In suitable preparations the callose plugs, which are present in the pollen tubes at regular intervals, fluoresce yellow to yellow/green colour.

9.2   Osmotic pressure and incompatibility in Linum grandiflorium.

Seeds of this highly ornamental annual plant can be obtained from most seed firms; they will flower 3–4 months after sowing and will produce plants of two types; half long-styled and half short-styled; the anthers are the same length in both types. Mount the pollen grains in the acid fucsin light green stain. Note that, unlike the pollen in the short- and long-styled primrose, they are of equal size. Make a series of solutions of sucrose from 5% to 80% in steps of five. Place the pollen and styles separately in the range of concentration and observe the bursting or plasmolysis of the pollen and the bending of the style. To show the difference between the two types of pollen it is necessary to use the whole range of concentration from 5–80%; for the styles the difference can best be observed by using 0–35% sucrose. The permeable stigmatic surface of the styles runs longtitudinally down one side of the style. In concentrations of sucrose which are isotonic with the cells of the stigmatic surface the style will remain straight and is therefore a measure of the suction pressure of the cells. Also self-pollinate and cross-pollinate two types, observe the pollen on the stigma and the bending of the stigma in the flower at hourly intervals after pollination. (Ref. LEWIS, D. (1943). Annals of Bot. N.S., VII 115–22.)

9.3   Pollen germination in sucrose

Pollen of many species of plants can be germinated in sucrose solution in a concentration between 5 and 25% by weight in water according to species. Some pollen germinates readily in the appropriate concentration but the tubes burst at the tip. The addition of borate in the form of sodium borate at .01% concentration usually prevents this bursting and assists the growth of the tubes. There are several simple methods of mounting the pollen in the solution for observation under the microscope. A drop of the solution is placed on a cover glass to which a few hundred pollen grains have been added by touching the surface of the solution with a dehisced anther. A glass or metal ring is attached to a microscope slide with a suitable waterproof adhesive and a few drops of water are placed in the bottom of the ring. The top of the ring is smeared with vaseline, and the cover glass with the drop of solution inverted quickly to prevent running of the drop and this is placed on the top of the vaseline ring. It can be watched for germination and growth under a microscope giving 50–150 magnification. Another method is to use a 1% agar base or a piece of filter paper for the sucrose solution placed on a microscope slide and the slide kept in a Petri dish with a pad of moistened paper to prevent drying out. The slide can be temporarily removed from the Petri dish for observation.

Further Reading

DARWIN, C. (1877). *The Different Forms of Flowers on Plants of the same Species.* (2nd Edition) John Murray, London.

HESLOP-HARRISON, J. and LEWIS, D. (1974). A discussion on incompatibility in flowering plants. *Proc. Roy. Soc. Lond., B.,* **188**, 233–375.

HESLOP-HARRISON, J. (1978). *Cellular Recognition Systems in Plants.* Studies in Biology, no. 100. Edward Arnold, London.

KNOX, R. B. (1979). *Pollen and Allergy.* Studies in Biology, no. 107. Edward Arnold, London.

LEWIS, D. (1976). Incompatibility in flowering plants. *Receptors and Recognition Series A,* Volume 2. Eds. P. Cuatrecasas and M. F. Greaves. Chapman Hall, London.

DE NETTANCOURT, D. (1977). *Incompatibility in Angiosperms.* Springer, Verlag, Berlin, pp. 1–223.

Appendix: Types of Incompatibility

	Genes	Alleles	Allelic interaction	Stage of inhibition	Effect of polyploidy	Type of pollen	Families
Heteromorphic							
1 Distyly	1	2	Dominant	Germination, Penetration, Pollen tube	None	Binucleate	Primulaceae, Lythraceae, Oxalidaceae, Plumb... 14
2 Tristyly	2	2	Dominant	Pollen tube	None	Binucleate	Lythraceae, P... Oxalidaceae 3
3 Distyly / Tristyly spurious	1	Many	Codominant	Fertilization		Binucleate	Narcis... 1
Homomorphic							
4 Gametophytic	1	Many	Codominant	Pollen tube	Breakdown	Binucleate	
5 Gametophytic	2	Many	Codominant	Pollen tube	None	Trinucleate	
6 Gametophytic	4	Many	Codominant	Pollen tube	None	Binuc...	
7 Sporophytic	1	Many	Dominant Codominant	Germination, Penetration	None	Tr...	
8 Sporophytic	3–4	Many	Dominant Codominant	Germination, Penetration	None		
9 Sporophytic Gametophytic	1	Many	Codominant	Fertilization			
10 Polygenic	Many	?					
11 Complementary Gametophytic	3–4	Many	Codominant				